HMS *Victorious* early in her long career. *(Author's Collection)*

HMS VICTORIOUS
1937 – 1969

Neil McCart

FOREWORD BY
VICE-ADMIRAL SIR IAN McINTOSH KBE DSO DSC

To All Those Who Served In HMS *Victorious* 1941-1968

Front Cover: A magnificent watercolour painting of HMS *Victorious* by artist Brian Conroy, Greatham, Hampshire.

Cover Design by Louise McCart
© Neil McCart/FAN PUBLICATIONS 1998
ISBN: 1 901225 01 1

Typesetting By: Highlight Type Bureau Ltd,
Clifton House, 2 Clifton Villas,
Bradford, West Yorkshire BD8 7BY

Printing By: The Amadeus Press Ltd,
517 Leeds Road, Huddersfield,
West Yorkshire HD2 1YJ

Published By FAN PUBLICATIONS
17 Wymans Lane, Cheltenham, GL51 9QA, England. Fax & Tel 01242 580290

Contents

Page No.

Foreword by Vice-Admiral Sir Ian McIntosh KBE DSO DSC 4

Chapter One ....Europe Prepares For War 5
Chapter Two ....Action With The *Bismarck*. 11
Chapter Three ......Northern Waters 19
Chapter Four ....'Operation Pedestal' 35
Chapter Five ....'Operation Torch' And The Pacific 44
Chapter Six ....'Operation Tungsten' And *Tirpitz*. 55
Chapter Seven ......East Of Suez 62
Chapter Eight ......The 'Meridian' Operations 70
Chapter Nine ......Return To The Pacific 78
Chapter Ten ....Troopin' And Training 91
Chapter Eleven ....Almost A New Ship 105
Chapter Twelve ....Crisis In The Persian Gulf. 120
Chapter Thirteen ....Two Commissions In The Far East 138
Chapter Fourteen ....The Final Commission 158
Chapter Fifteen ....The End Of The Story 176
Appendix One ......Principal Particulars 182
Appendix Two ......Commanding Officers 183
Appendix Three ....Previous Ships 184
Appendix Four ......Rules For Passengers 185
Appendix Five ......Battle Honours 189
Appendix Six ......Aircraft Types HMS *Victorious* 190

Foreword

HMS *Victorious*, in her 27 years of service in the Royal Navy, had a life of such variety and distinction as few of HM ships could equal.

In action shortly after she first commissioned, in the North Atlantic at the sinking of the powerful German battleship *Bismarck*, and later against that ship's sister *Tirpitz* in Arctic waters where she also provided help for the grim Murmansk convoys attacking airfields and destroying aircraft threatening them. She served in the Pacific with the US Fleet after Pearl Harbor and later, in the Mediterranean, protected vital convoys relieving Malta and supported the Allied landings in North Africa. In South-East Asia she struck crippling blows against the Japanese with her attacks on oil refineries and storage tanks. In the British Pacific Fleet she supported the US taking of Okinawa and in the preparation for the final assault on the Japanese heartland. She survived hits by kamikaze aircraft, operating her aircraft again within hours.

With the end of what for her was a very active war, her peacetime duties were varied and wide-ranging covering humanitarian and peace-keeping, and peace-enforcing, activities across much of the world. Indeed in war and peace it can be said that she was, in Richard Hakluyt's words, 'compassing the vast globe of the earth'.

The reader of this admirably researched account which is both detailed and succinct will be saddened to read of the loss of so many young aviators in peace as well as in war. This does highlight the fact that flying from an aircraft carrier, even in what always seemed to be rare favourable weather conditions, is a dangerous business and this country is fortunate that in the Fleet Air Arm it has young men of courage and outstanding skills to carry out the tasks which contribute so much to the achievement and maintenance of peace. I should add that I have never come across any aircrew who would wish to change - not even to become one of their three-dimensional colleagues, the submariners.

This is a book not only for those who had the good fortune to serve in this great ship or for those who have a particular interest in the history of the Royal Navy, much pleasure and information though they will gain, but I commend it also to that wider readership for whom the dedication and gallantry of men are counted as qualities worthy of praise and emulation.

Ian McIntosh KBE DSO DSC
Vice-Admiral

Chapter One

Europe Prepares For War

The year was 1936. It saw the giant Cunard liner *Queen Mary* make her maiden voyage from Southampton to New York and the first television broadcasts from the BBC's Alexandra Palace Studios, but on a more sombre note, it was also the year in which the great authors Rudyard Kipling and G. K. Chesterton died, the Jarrow hunger marches began and King Edward VIII abdicated the throne after a reign of only 325 days. For the continent of Europe as a whole there were ominous signs of a German military resurgence and a determination to repudiate the Treaties of Versailles and Locarno, which most Germans, and for that matter many British people, considered grossly unfair. The first indication of trouble ahead came in March that year with the German remilitarization of the Rhineland.

It was at this time that the London Naval Agreement was signed with Britain, France and the United States approving a 35,000-ton limitation on new warships with restrictions on the size of guns to be fitted to the vessels. This was all very well, but the Italian and Japanese Governments refused to sign the final agreement and in Germany construction began on the powerful 45,000-ton battleships *Bismarck* and *Tirpitz*. However, the agreement suited Britain for with a weak economy the Government had only a limited budget for defence expenditure. The first post-war naval conference, which had been held in Washington from November 1921 to February 1922, dictated that only a limited number of new warships should be built and stipulated that 20,000 tons be the maximum for new aircraft carriers. This was perhaps the

On 14 September 1939, just after the British and French declaration of war, HMS *Victorious* was launched on the River Tyne. *(Fleet Air Arm Museum)*

An aerial view of the *Victorious* taken by the RAF on 16 April 1941, as she steamed from the River Tyne to Rosyth. She is under way at full power, and her camouflage markings on the flight deck stand out clearly on what was a fine day.
(Public Record Office AIR 28/828. Crown Copyright)

single most important factor in the Royal Navy's construction programme during the 1920s and early 1930s, when the service relied on small and inadequate aircraft carriers which were mainly converted from hulls originally intended for other types of vessel. Although the years following the Great War saw rapid advances in the development of aviation, little of this progress was apparent in the Royal Navy which remained, in aviation terms, the poor relation of the newly formed Royal Air Force, particularly as far as the production of fast modern aircraft was concerned. Fortunately, in 1935 the decision was made to build a large new aircraft carrier, which eventually became the *Ark Royal*, and in April 1936, as part of that year's Defence Estimates, the Government announced plans to build 38 new warships, including four aircraft carriers of the Illustrious class which were to be based on the design of the *Ark Royal*, but which would embody many improvements on that ship.

The new carriers were to be named *Illustrious*, *Victorious*, *Formidable* and *Indefatigable*, and they were to have a displacement of 23,000 tons, an overall length of 753ft and a beam of 95ft, which would allow them to use dry docks in the Naval Dockyards at Portsmouth, Devonport and Rosyth, as well as the docks in Malta and Singapore. The limitation on tonnage had been set by the London Treaty and, as a result, the two-level hangars which had been incorporated into the *Ark Royal's* design had to give way to just one level. The hangars themselves were to be built with side armour 4.5 inches thick and together with a 3-inch armoured deck and deckhead this would create an armoured citadel which would give protection against 500lb bombs dropped from up to 7,000ft. The side armour would protect against 6-inch shells, while underwater the whole length of the armoured citadel was designed to withstand attack by torpedoes carrying a charge of 750lbs. The aircraft hangars would be 456ft long and 62ft wide with a height of 16ft and would accommodate 30 torpedo, spotter and reconnaissance aircraft, and six fighters. Electrically operated lifts, each being 45ft long and 22ft wide, were to be positioned at each end of the hangar, and provision was made for one hydraulic accelerator, six arresting gear wires, two electric cranes, a transverse windscreen before the forward lift and side windscreens abreast the accelerator.

The vessels would have triple-screw machinery capable of developing 111,000 SHP, and this would give the ships a top speed of just over 30 knots. The machinery was to be arranged in three engine rooms and three boiler rooms abreast, each boiler room having two boilers, and the engine rooms were to be 36ft apart to give some protection against underwater attack. Stowage was arranged for 4,640 tons of oil fuel, which would be sufficient to provide an endurance of 12,500 miles at 14 knots. Provision was also made for six, 400kW turbo-generators, two of which would be fitted in the wing engine rooms, with the remaining four being situated on the main deck within the armoured citadel.

Armament for the new carriers would consist of 16, 4.5-inch HA/LA guns in eight twin mountings, which would be grouped at the forward and after ends of the ships thus leaving the midships section of the flight deck available for parking aircraft free from excessive gun-blast. They would also be equipped with six multiple pom-poms, three on either side of the flight deck.

Accommodation was to be provided for a full complement of 1,286, which included provision for an Admiral and his staff. However, the Admiralty documents at the time stated that, 'The accommodation is congested and it has not been found practicable to provide recreation spaces or reading-rooms for CPOs and POs.'

The *Victorious* was the third ship of the class to be ordered and on 28 November 1936 the Admiralty requested tenders from ten shipbuilding companies, including John Brown & Co of Clydebank, Harland & Wolff Ltd of Belfast, Cammell Laird & Co of Birkenhead, who were actually building the *Ark Royal*, and Vickers Armstrong of both Barrow-in-Furness and Walker-upon-Tyne. Replies were required to be at the Admiralty offices in Bath by the morning of 4 January 1937, where they were all opened that same day. The highest tenders were from the Clyde shipyards of John Brown and Scotts who quoted some £129,000 higher than Vickers Armstrong of Walker-upon-Tyne, whose bid totalled £2,464,000. A breakdown of this figure showed £1,694,000 for the building of the hull, £702,000 for the main propulsion machinery and £68,000 for the auxiliary machinery. The lowest of the tenders came from Vickers Armstrong at Barrow which was exactly £1,000 cheaper than the sister company on the Tyne. It was some three months later, on 13 April 1937, that a decision was finally made by the Admiralty and in the following week an official order was sent to Vickers Armstrong at Walker-upon-Tyne, 'For the construction and completion in all respects of one aircraft carrier for His Majesty's Navy.' One aspect of the price which caused some interest in the Admiralty was the fact that the *Victorious* would cost £59,971 more than the *Ark Royal* which had been ordered less than two years previously.

The first keel plates for the new carrier were laid on 4 May 1937, just two weeks before Neville Chamberlain succeeded Stanley Baldwin as Prime Minister, and over the next two eventful years the hull took shape in the Tyneside shipyard. In March 1939, as German troops occupied the rump of Czechoslovakia, which had been left after Munich, the Admiralty announced that the *Victorious* would be launched in September that year. The actual date was set for 14 September which, in the event, was only eleven days after the British and French declaration of war against Germany, and the launching ceremony took place against the backdrop of the German Army's powerful offensive into Poland, their encirclement of the Polish capital Warsaw and the preparations of the Soviet Army to invade Poland's eastern frontier. Germany's repudiation of the Treaty of Versailles was complete and the 'unfinished business' of the Great War of 1914-18 was resumed.

Although the launching of the *Victorious* was carried out with all the traditional ceremony, strict press censorship meant that it received none of the publicity which is usual on such occasions. The new vessel's sponsor was Lady Augusta Helen Inskip, the daughter of the seventh Earl of Glasgow and the widow of Charles Orr-Ewing who had been the MP for Ayr. In 1914 she had married Thomas W. H. Inskip, a barrister, who became the MP for a Bristol constituency in 1918. During the Great War Thomas

Inskip had worked for Naval Intelligence in London where he had been head of the Naval Law Branch, and he represented the Admiralty on the War Crimes Committee of 1918. During the inter-war years he had served in Government as Attorney General, and in 1936 he was appointed as minister for the co-ordination of defence as the country started its rearmament programme. It was largely as a result of Inskip's work that, in 1937, the Air Ministry agreed to relinquish control of the Fleet Air Arm and the Admiralty quite rightly took full control of carrier-borne aircraft. In January 1939 Inskip was transferred to the Dominions Office and his successor, Admiral of the Fleet Lord Chatfield, went out of his way to tell the country that it was deeply indebted to Inskip in removing control of the Fleet Air Arm from the RAF in what had been a secret and, largely, a thankless task. In May 1940, having been created Viscount Caldecote, he returned to the Dominions Office and was leader of the House of Lords. In October 1940 he was appointed Lord Chief Justice and he will be remembered as a good judge who was courteous and patient, and sound in his conclusions. It was fitting, therefore, that Lady Inskip, as the wife of a man who fully understood the power and significance of carrier-borne aircraft, should launch one of the Royal Navy's latest and most powerful aircraft carriers.

The uncertainty of the times ensured that the usual excitement which accompanies such launching ceremonies was a little subdued, but the shipyard was crowded with the workers and their families and there were cheers from those assembled when, at noon on 14 September 1939, Lady Inskip sent the *Victorious* down the slipway and into the waters of the River Tyne, first having christened her with a bottle of the best champagne. As the drag chains brought the carrier's hull to a stop in midriver, the tugs commenced their work to bring her alongside the fitting-out wharf, the crowds dispersed and work was resumed in the shipyard. One rating who joined the *Victorious* on 15 September 1939 was Stoker Petty Officer Albert Buckley who recalls, 'She had just been towed upriver from the Wallsend slipway from where she had been launched, and I was accommodated in a furnished house in Walker-upon-Tyne until she was commissioned in March 1941. I was actually promoted to Chief Stoker in 1940, at the offices of Vickers Shipbuilders.'

It was just over a year after the launching ceremony, in October 1940, that the first key officers, including the carrier's first commanding officer, Captain Henry C. Bovell CBE RN, were appointed to the ship, and by the end of that year the fitting out of the *Victorious* was nearing completion. Captain Bovell had joined the Navy as a cadet in the early years of the 20th century and he had attended the Naval Colleges of Osborne on the Isle of Wight, and Dartmouth. His first ship, which he joined as a midshipman in 1910, was the battleship HMS *Lord Nelson*.

After being promoted to sub lieutenant in 1913 he specialized in gunnery and was promoted to captain in 1934.

One member of the ship's company who joined the *Victorious* in January 1941 was Leading Stores Assistant Les Vancura, who recalls those early days when, after a German air raid, the carrier was moved downriver while disposal teams dug for an unexploded bomb which had buried itself in the dockside at the base of a crane which was close to the ship. Les was put in charge of the ship's No 5 store and the Fleet Air Arm Maintenance Store, and over the following weeks he supervised the embarkation of stores. Les noted in his diary that the ship was to have been commissioned on 11 March 1941, but this was postponed until later in the month. Another member of the ship's company remembers that, 'When we first joined the ship at the Walker yard not one of us had ever been close to an aircraft, and our first machine was an Albacore which "flew" on board by means of a dockyard crane. It had no airscrew and very flat tyres and we learned how to push, where it was safe to do so, and how to fold and spread its wings. During the weeks alongside we continually pushed it along the flight deck, down the forward lift, through the hangar and back up to the flight deck on the after lift. For days on end we heaved and sweated and we grew to hate the machine, yet 50 years later I look back on it with affection.'

It was in the last week of March 1941 that the main body of the ship's company joined the *Victorious* and Charles Minett was one of those who left Devonport Barracks for the long, slow journey north, most of which was during the night. Charles remembers the day well: 'We eventually arrived at York railway station the following morning where we changed trains. The second train was literally made up of dining cars and the reason for this soon became clear for, as we pulled slowly out of York, waiters appeared and we were all personally served with a full breakfast, which was very nice and not the sort of service we were used to in the Navy. Eventually, that afternoon, we arrived at Walker-upon-Tyne railway station where the Royal Marines Band was waiting for us. We then formed up in threes and with the band leading the way we marched to the ship. As we left the vicinity of the railway station we marched over the brow of a hill, and there below us was the *Victorious* lying alongside the quay. It was my first sight of the ship which would become my home for almost five years and which would carry me right round the world to almost all theatres of war. Once at the jetty and alongside the ship we embarked and were shown to the hangar where, once again, we formed up into threes before being detailed to return to the quayside to carry a kitbag, not our own but the first one which we could grab, back on board and into the hangar. I can remember we were then led round what appeared to be a maze of passageways to our respective mess decks before being told to return to

Another view of the new carrier as she leaves the Tyne for the first time for Rosyth.
(Public Record Office AIR 28/828. Crown Copyright)

the hangar to collect our own kitbags. Having, with some difficulty, found my way back I eventually found my kitbag amongst the huge mound which had been brought on board, but then I realized that I had forgotten the way back to my mess deck. Fortunately, I soon found my way around the *Victorious*.'

A somewhat unusual sight on board were the RAF personnel and one RAF stores assistant recalls that, 'Although I had a kitbag, I had no hammock, but I managed to obtain one and then I had the pleasure of a crash course on how to assemble and hang the same plus, of course, the correct way to get into it and, more importantly, stay in it. I remember thinking, "What am I doing here? I joined the RAF not the Navy." However, I did enjoy being an RAF Penguin.'

By the end of March 1941, when the *Victorious* was ready for commissioning, the war situation was very different from that which had prevailed in the autumn of 1939 when Lady Inskip had launched the carrier. Germany dominated most of Western Europe and Italy had joined the Axis cause, which made the Mediterranean a very hostile area. In Britain the economy was virtually bankrupt, but fortunately President Roosevelt had managed to amend the United States' Neutrality Act in order to permit the sale of arms and equipment and he had requested that Congress grant seven billion dollars in military credits to Britain. Although it was not generally known at the time, salvation lay in the fact that Hitler and his generals were concentrating all their country's military planning and effort towards the massive offensive which was code-named 'Operation Barbarossa', the invasion of Russia. The 18 months between September 1939 and March 1941 had seen the Royal Navy engaged in some significant sea battles, notably off the River Plate where three British cruisers, which were outranged and outgunned, fought a brilliant action against the pocket battleship *Admiral Graf Spee* and forced her to retreat to the safety of the neutral port of Montevideo where she subsequently scuttled herself. The Navy's air attack on the Italian Naval Base of Taranto showed to the world for the first time the awesome power of maritime aviation. On the day before the *Victorious* was commissioned the Battle of Matapan was fought in the Mediterranean, which proved to be another victory for the Royal Navy.

It was at 11.30am on Saturday 29 March 1941, at the Vickers Armstrong shipyard, that HMS *Victorious* was commissioned with a full complement, excluding squadron personnel, under the command of Captain Bovell, although it would be another 18 days before she was ready to put to sea for the first time. During these final days at the shipyard, air raid warnings were frequent and the ship's pom-pom crews and air lookouts were continually at their stations, particularly in the hours just before and after midnight. Two weeks after commissioning, during the afternoon of Saturday 12 April, the First Lord of the Admiralty, Mr A. V. Alexander, visited the carrier and spoke to the ship's company. Three days later the ship was prepared for sea, and at 5.50am on the morning of Wednesday 16 April the *Victorious* slipped her mooring ropes and set course down the River Tyne for the North Sea. Boy Seaman Harry Blake remembers, 'When we sailed downriver it was a very pleasant day and we were all closed up at Action Stations. I was closed up at the navigational rangefinder and the Officer of the Watch kept us busy for quite some time.' Just over an hour after leaving her berth the *Victorious* passed Tynemouth breakwater and the tugs which had guided her downriver were slipped before, with all hands at their stations, and air lookouts and pom-pom crews closed up, the carrier set course for Rosyth Dockyard and the engineers were able to work the engines up to full speed for the first time.

As soon as she passed through the harbour entrance the *Victorious* was joined by an escort screen which included the cruiser *Naiad* and the Hunt-class destroyers *Quantock* and *Southdown*, whilst overhead the RAF provided a fighter escort for the ten-hour voyage through the dangerous waters of the North Sea. Fortunately, the carrier's passage was without incident, and at 5.15pm that same afternoon she anchored safely off Rosyth Naval Dockyard. Soon after this the guns' crews and lookouts were stood down and everyone was able to relax once again. Next day, at 5.30pm, the *Victorious* was manoeuvred into No 1 dry dock in Rosyth Dockyard where the underwater hull could be scraped and repainted and, more importantly, many members of the ship's company were able to take some leave.

Chapter Two

Action With The *Bismarck*

During her stay at Rosyth ammunition and aviation fuel were embarked, and on Sunday 4 May 1941 the dry dock was flooded for the *Victorious*' departure, and that evening she anchored in midriver once again. Meanwhile, in the Admiralty it had been decided that the *Victorious*' first operational role would be that of an aircraft ferry and that she would deliver 50 RAF Hurricane fighters to the beleaguered British garrison in Egypt. Initially the reinforcement aircraft were shipped to Takoradi in West Africa before flying almost 4,000 miles to Egypt, via Khartoum in Sudan. However, another idea put forward was to ferry the aircraft in aircraft carriers from the UK into the Mediterranean, where they would be launched for Malta before continuing to the Western Desert. One RAF officer who was involved in these operations was Group Captain H. F. O'Neill DFC who, as a young pilot officer, had previously made the flight to Egypt by way of Takoradi and in April 1941 he received orders to travel from his station in Hampshire to Rosyth to join an aircraft carrier bound for the Mediterranean where he would fly off the carrier and make the flight to Malta and on to Egypt. Meanwhile, his squadron's ground parties were to go by sea to Cape Town and then by rail and air to Cairo.

Back at Rosyth the *Victorious* had been moved into the dockyard basin where she was secured alongside the West Wall, and for five days from Friday 9 May to Tuesday 13 May, 48 crated Hurricanes were embarked by the Fleet Air Arm personnel. Two days later, at just after 5.30am on Thursday 15 May, the *Victorious* left Rosyth for trials off Burntisland and then for the voyage north to Scapa Flow, where she anchored at just before 11pm that evening. Meanwhile, back in Hampshire Pilot Officer O'Neill and his colleagues were preparing for their journey north and he remembers, 'A few wives and girlfriends had formed up on Andover station to see us off on the slow journey to Scotland. A second Hurricane squadron joined us in Edinburgh, where we were surprised to find that our carrier had sailed from Rosyth for Scapa Flow. So we

Captain Bovell inspecting Divisions on the flight deck whilst the ship was at Scapa Flow. *(C. Minett)*

The *Victorious* steams through choppy seas on 24 May 1941 as she closes her flying-off position so that nine Swordfish of 825 Squadron, seen here on the after end of her flight deck, could attack the German battleship *Bismarck*. (S. Filmer)

puffed our way up to the "spacious melancholy of Caithness", before coming to rest in the tiny port of Scrabster. Casting off from there, as generations of the Navy had done before us, we bore down on our temporary home for the next few weeks - the aircraft carrier *Victorious*, lying silent and grey in the Flow with other ships of the Home Fleet. We must have seemed a pretty curious lot to those of the ship's company who had assembled to welcome us. *Victorious* was a new ship and was still working up, so our arrival on board must have been a headache for Captain Bovell and his executive officers. Nevertheless, they put themselves out to make us feel very much at home in the time-honoured tradition of the Royal Navy.' Group Captain O' Neill goes on to explain what happened to his squadron's Hurricanes: 'The aircraft lifts in the ship were too small to accommodate an assembled Hurricane, so our aircraft had already been taken apart and stowed in the hangar with engines "inhibited" while not in use. We were therefore purely passengers on board and did our best to keep out of the way of those who had serious work to do. Flight Lieutenant E. J. Morris and I shared a cabin aft over

the thundering screws and its number - 109 - was an ominous reminder that we were on our way to some serious business and not in the ship for a luxury cruise.'

It was intended that the *Victorious* would sail from the Clyde on Thursday 22 May 1941, in company with the fast troop convoy WS8B and the battlecruiser *Repulse*. In the meantime there were a few days which the *Victorious* could use to continue her short work-up in the Pentland Firth, and she left the Flow on Saturday 17 May to carry out gunnery exercises and full-power trials off Dunnett Head. On Sunday 18 May, after an early start, and escorted by the destroyers *Achates*, *Icarus* and *Active*, the *Victorious* undertook her first flying operations when, at 9am that morning, a flight of six Fulmars from RNAS Hatston carried out landing and launching trials. They were followed by Swordfish aircraft of 825 Naval Air Squadron and Sea Hurricanes, and that afternoon a Fairey Barracuda, which was on loan to 778 Squadron, carried out deck landing trials. Its arrival is recalled here by Group Captain O' Neill: 'The island superstructure was our vantage point for most of the voyage and it was from there that we watched the bows go round as the ship turned into the wind for flying-on, leaving a curving wake of cream and green that shimmered into the hulls of the destroyer escorts. That remarkable naval aeroplane the Barracuda made its debut on the flight deck and all were startled at the spectacle of such an extraordinary aerial device slow-rolling at low altitude with a torpedo attached to it.'

The Fairey Barracuda was a very odd looking torpedo bomber which had originally been ordered in July 1938 with the prototype being flown in December 1940. However, problems arose with the low-set tailplane and by the time this had been redesigned it was 1942 before the aircraft was completed, priority having been given to RAF aircraft, and so the plane which carried out trials on *Victorious* had the old tail unit fitted.

That evening the *Victorious* anchored in Scapa Flow and next morning there was another early start as she left the area to continue with her flying trials. The Barracuda repeated the deck landing trials, before returning to the Fairey Aircraft Company for a new design of tail to be fitted. After landing on the Swordfish aircraft of 825 Squadron the carrier returned to Scapa Flow, where she anchored at 6.15pm that evening. Les Vancura recalls that the ship's company were issued with tropical kit and all the civilian contractors left the ship at that time, which led to numerous rumours about the ship's destination.

However, unknown to most of those on board the *Victorious*, the powerful German battleship *Bismarck*, which Hitler had described only days before as the 'Pride of the German Navy', together with the heavy cruiser *Prinz Eugen*, had left Gdynia during the evening of Sunday 18 May and the intelligence reports received by the C-in-C Home Fleet, Admiral Sir John Tovey, indicated that the two vessels were about to break out into the Atlantic. The two German warships were actually setting out on 'Operation Rheinburg', whose objectives were to launch raids on British convoys on the Atlantic, and Admiral Raeder would have preferred to send a larger force including the *Scharnhorst* and *Gneisenau*. Both were in the French port of Brest following a devastating series of raids on convoys off Newfoundland, but the former was under refit and the latter had been damaged in an air raid, so for the time being the *Bismarck* and *Prinz Eugen* would carry out the raids which were intended to cause havoc to vital Atlantic convoys. However, in order to break through into the Atlantic Ocean the two German ships would have to pass through the waters of the Norwegian Sea, between the Orkney Islands and Iceland, or through the Denmark Strait between Iceland and Greenland. The first route was tightly patrolled by the Royal Navy from the Home Fleet's base at Scapa Flow and the latter route was only navigable during the summer months, and even then with the frozen waters and thick banks of fog and mist it would be a hazardous passage. The first reports that the two German ships were at sea came from the Swedish Navy cruiser *Gotland*, after which it became clear that they had anchored off Bergen.

At 11am on Tuesday 20 May, Admiral Sir John Tovey visited his one aircraft carrier at Scapa Flow, the brand new *Victorious*, where he addressed the ship's company. During the remainder of that day the carrier refuelled and empty shell cases were discharged into a lighter. At about 11pm on Wednesday 21 May, unknown to Admiral Tovey, the *Bismarck* and *Prinz Eugen* left Bergen to make their passage through the Denmark Strait. However, the Admiral was well aware that the enemy ships could leave at any time and at ten minutes after midnight on the morning of Thursday 22 May, he ordered his Battle Cruiser Force of HMS *Hood* and HMS *Prince of Wales* and the destroyers *Electra*, *Echo*, *Anthony*, *Icarus*, *Achates* and *Antelope* to leave Scapa Flow for Seydisfjord in Iceland where they would be in a better position to cover the Denmark Strait and the Iceland Faeroes gap in the Norwegian Sea. During the day there was intensive air reconnaissance over Bergen, but with a low mist hanging over the area it was early evening before it became clear that the German ships had left their anchorage. The *Victorious* had spent the day at sea with the destroyers *Windsor*, *Ferndale* and *Lance*, carrying out deck landing practice with Swordfish and Fulmar aircraft. It was almost 8pm when she returned to her anchorage at Scapa, but it was not for long for at 10.20pm she weighed anchor and put to sea with the battleship *King George V*, flying the flag of Admiral Sir John Tovey, the cruisers *Aurora*, *Galatea*, *Kenya* and *Hermione*, and seven destroyers. It was at just after 11pm when the force passed Hoxa Boom with the *Victorious* in station astern of the flagship. At 2am on Friday 23 May the force was off Cape Wrath, and at noon that day the *Bismarck* and *Prinz Eugen* entered the narrowest part of the Denmark Strait where, at 7.22pm, they were sighted by the cruiser *Suffolk* which was patrolling the area with her sister ship *Norfolk*. The *Suffolk* steamed into a nearby mist bank where she was able to shadow the German ships in relative safety. However, about an hour later when the *Norfolk* joined her sister, the *Bismarck* opened fire on her at a range of about six miles and, although none of the five salvoes hit the cruiser, the two ships subsequently shadowed the German warships from a safe distance. During the morning of 23 May Admiral Tovey's force was joined by the battlecruiser *Repulse* and two more destroyers, and that night the shadowing cruisers had a difficult time keeping in contact because of heavy snowstorms. By 4am on Saturday 24 May Admiral Lancelot Holland in HMS *Hood* was only 20 miles south-east of the two German ships and soon after this he ordered a change of course to close the range between his force and the enemy ships. At about 5am the *Norfolk* and *Suffolk* sighted the *Hood* and *Prince of Wales* and at 5.52am the *Hood* opened fire at a range of about 26,500 yards. Within minutes the German ships replied

A Swordfish of 825 Squadron, armed with a torpedo, takes off from the *Victorious* on the morning of Sunday 25 May 1941, in one of the follow-up searches to attack the *Bismarck*. (*S. Filmer*)

HMS *Victorious* at anchor in Scapa Flow. Note how the wartime censors have blanked out the ship's radar aerials. (*S. Filmer*)

and although the *Bismarck's* first two salvoes were wide, her third salvo straddled the *Hood* and a salvo from the *Prinz Eugen* hit the elderly battlecruiser and started a fierce fire amidships. At 6am, with smoke pouring from the fires, a 15-inch shell from the *Bismarck's* fourth salvo plunged through the *Hood's* thin armoured deck and detonated in her 4-inch magazine which then ignited the main magazine with the result that the *Hood* was torn in half by a massive explosion. Within a few minutes both the forepart and the stern had risen vertically in the water shrouded by an enormous cloud of black smoke, and then both sections sank beneath the waves taking with them all but three of her 95 officers and 1,324 men.

Immediately the *Prince of Wales* had to turn sharply to starboard to avoid the wreckage of the *Hood*, and then she in turn came under the concentrated fire of both German warships. She was hit seven times, with one 15-inch shell crashing through into the bridge where, although it did not explode, in the resulting chaos all the personnel apart from Captain J. Leach and the Chief Signals Yeoman were killed or injured by the flying splinters and fragments of debris. The ship was suffering continual problems with her main armament, and being no match for the combined fire of the two German ships Captain Leach broke off the engagement and retired under cover of a smokescreen. Although his action in withdrawing the *Prince of Wales* caused some controversy in Admiralty circles, it was the only sensible thing to do for the ship's after 14-inch turret was jammed and only B turret was completely operational. Although it was not known at the time, the *Prince of Wales* had scored three hits on the *Bismarck*, two of which had caused serious damage to the battleship. One had damaged a boiler room and had put two boilers and a turbo-generator out of action, and the other, although it had not exploded, had passed through the ship and seriously damaged fuel tanks and pumping machinery. This left her with a list to port, and it could be seen that she was leaking heavy furnace fuel oil in her wake. Despite this, the first engagement with the *Bismarck* had ended in disaster for the Royal Navy.

Meanwhile, in a position Lat 61° - 15'N/Long 22° - 08'W, over 300 miles from the *Bismarck* and *Prinz Eugen*, Admiral Tovey in *King George V* decided that his best chance of reducing the German battleship's speed was to launch an aerial attack from the *Victorious*, despite the fact that she was not fully worked up. On board the carrier nine Swordfish aircraft of 825 Squadron, commanded by Lt-Cdr Eugene Esmonde, had been embarked, and at 4pm Admiral Tovey ordered the *Victorious*, together with a cruiser screen consisting of the *Galatea*, *Aurora*, *Kenya* and *Hermione* under the command of Rear-Admiral A. T. B. Curteis, to part company and to steam hard to a position approximately 100 miles from the two enemy ships. For over six hours the carrier, together with her screen of cruisers, battled through heavy seas and driving rain at 30 knots until, at 10pm, Admiral Curteis decided he was close enough to launch an air strike.

Group Captain O' Neill recalls that: 'We in the two Hurricane squadrons took no part in the affair because our aeroplanes were stowed in pieces in the hangar. At one point there was a move to assemble some and range them on deck, but fortunately the idea was swiftly discarded. It is impossible to imagine how we could have contributed to the action - other than in a reconnaissance role - and we should have been thoroughly in the way of the Fleet Air Arm operations. Our stowed aircraft survived the tremendous seas as we pounded along south of Iceland at 30 knots giving a rise and fall to the flight deck of some 60 feet. We did our best to keep out of the way of everyone else and gave moral support to the Fleet Air Arm crews, who were operating in filthy weather and whose professional skill and courage in the action were quite unbelievable. There were few who were not stirred by the sight of Esmonde and his observer Lt Ennever quietly slipping away first in their Swordfish, from a rolling and pitching flight deck.'

Despite the heaving flight deck, the howling wind, the rain and the spray, the nine Swordfish, each armed with an 18-inch torpedo, took off at 10.14pm on Saturday 24 May from the carrier and set course for the *Bismarck*, aided in their search by a very basic radar set. In spite of the foul weather conditions and the fact that the *Victorious* was not fully worked up, all nine aircraft were airborne within five minutes of the first launch, and at 10.55pm three Fulmars were launched with orders to shadow the attacking force. It was at just before 11.50pm that the Swordfish thought they had made contact with the *Bismarck*, which, in fact, turned out to be the US Coastguard cutter *Madoc*, but guided by the shadowing cruisers *Norfolk* and *Suffolk*, it was not long before they located the two German warships. Despite heavy anti-aircraft fire from the *Bismarck*, all but one of the Swordfish pressed home their attack and one of them scored a hit on the battleship's starboard side amidships which an observer in one of the shadowing Fulmars described as, '...a great, black column of dense smoke rising from the starboard side.' On board the *Bismarck* a 'tremendous shudder' ran through the ship forcing out a towering column of water which rose from the starboard side. Although the German battleship had sustained only one hit this time, it was critical for it compounded the damage in the boiler room which had been caused by the *Prince of Wales*, with the result that the *Bismarck's* speed was temporarily reduced to 16 knots. At 1.10am two more Fulmars were flown off to relieve the first aircraft, but 18 minutes later only one of the original three Fulmars returned to the *Victorious*. The pilot had to land on a flight deck which was rolling and pitching violently as the carrier steamed into winds of 40 knots, with high waves shrouded

The *Victorious* approaches Gibraltar with Hurricanes on the flight deck. *(Fleet Air Arm Museum)*

in dense streaks of foam and much reduced visibility because of the driving rainstorms. So bad were the weather conditions that down below deadlights and scuttles had sprung open and mess decks were awash with sea water, which kept the shipwrights busy. At 2am on the morning of Sunday 25 May, with the weather conditions having deteriorated even further, the nine Swordfish aircraft arrived back over the *Victorious* and as the ship's bridge log succinctly records: '02.00: Commenced landing on nine Swordfish and one Fulmar. 02.25: Landing on completed.' These two brief sentences do not do justice to a remarkable achievement by all the aircrew and the new carrier's ship's company. Fortunately, the crews of the shadowing Fulmars which had been forced to ditch were rescued by a passing merchant ship.

Shortly after the air attack the shadowing cruisers *Norfolk* and *Suffolk* lost contact with the German warships, but Force II which was based at Gibraltar and included the aircraft carrier *Ark Royal*, were steaming north and at 7.50am, less than six hours after they had landed on, seven Swordfish were airborne and carrying out radial searches for the *Bismarck* and *Prinz Eugen*. By this time the *Victorious* was in a position Lat 56° - 18'N/Long 36° - 29'W which was, in fact, because of a course change by the German ships, which had put them well out of range of the Swordfish aircraft of 825 Squadron. However, at just after 11am, at the end of their mission, there were only six Swordfish waiting to land on once again. The seventh aircraft, which was being flown by Lt Pat Jackson, with his observer Sub Lt Berrill and their Air Gunner L/Tel Sparkes, had been unable to find the carrier and had ditched south-east of Iceland. Despite Force 6 winds and a sea where large waves topped with white foam crests had formed, Lt Jackson managed to bring the Swordfish down near to an abandoned lifeboat. All three men were able to climb into the almost waterlogged craft, which they had to bale out repeatedly with their flying boots. The lifeboat had come from the Dutch oil tanker *Elusa* which had been torpedoed in the North Atlantic on 21 May. On the ninth day of their ordeal, in driving snow and at the end of their tether, they managed to attract the attention of the Icelandic grain carrier *Lagarfoss*, which picked them up and took them to Reykjavik where they learned of the *Bismarck's* ultimate fate.

Meanwhile, back on board the *Victorious* the searches for *Bismarck* continued throughout Sunday 25 May, with the last aircraft landing on at midnight. Next morning search operations began at 5am and at 12.50pm the first Swordfish strike was flown off the *Ark Royal* which was in a position Lat 48° - 14'N/Long 19° - 15'W. This sortie almost ended in disaster when the aircraft crew mistook the cruiser HMS *Suffolk* for the German battleship and launched their attack. Skilful manoeuvring by Captain Larcom in the cruiser enabled the vessel to evade the torpedoes and that evening a second strike by the *Ark's* aircraft succeeded in crippling the *Bismarck's* steering gear. This was the end for the mighty German warship and at 10.40am on the next day she was sunk by units of the Home Fleet, including the *Rodney*, with the *coup de grâce* being a torpedo fired by HMS *Dorsetshire*. The tragedy of the *Hood* had been avenged, but this time the long roll of the dead and missing belonged to the Kriegsmarine.

On Tuesday 27 May the *Victorious* set course for the Clyde and she reached her destination two days later, anchoring off Greenock at 2pm on Thursday 29 May. Les Vancura has recorded in his diary that the ship was cheered as she made her way into the River Clyde, so her exploits and her role in the sinking of the *Bismarck* had obviously been well publicized. Next day the three Swordfish and three Fulmar aircraft which had been lost were replaced and the last day of May, which Les Vancura has recorded as, 'a real scorcher', was spent embarking stores from lighters. However, the welcome break at anchor would not last for long, and at 10.30pm that same evening she left to join convoy W58X bound for Gibraltar. Soon after leaving the Clyde a memorial service was held for the aircrew who had been lost during operations against the *Bismarck*. Also in the escort were the cruisers *Norfolk* and *Neptune*, and the destroyers *Legion*, *Wivern*, *Wild Swan*, *Sherwood*, *St Mary's* and *Brighton*, the latter three being ex-US Navy destroyers which had been transferred to the Royal Navy under the Lend-Lease agreements. Included among the merchantmen were the *Port Wyndham*, the *Esperance Bay* and the 20,000-ton Canadian Pacific liner *Duchess of Bedford*. The convoy was routed well out into the Atlantic and by midday on Tuesday 3 June they were in a position Lat 49° - 04'N/Long 20° - 20'W, with the weather much warmer now as Group Captain O' Neill recalls: 'I spent some pleasant moments during the voyage perched in the sunshine on the round down of the flight deck, enjoying the view over the receding wake.' During the voyage reconnaissance patrols were flown continually and on Wednesday 4 June a Swordfish of 825 Squadron sighted the German supply-cum-patrol ship, *Gonzenheim*, about 100 miles north of the Azores. She had been scheduled to rendezvous with the *Bismarck* and the fact that the cruiser *Neptune* had been dispatched to intercept her, persuaded the German captain to scuttle his ship. In the event her crew of 63 were picked up by the cruiser. By the morning of Monday 9 June the *Victorious* was nearing Gibraltar and at 6.50am the *Ark Royal*, *Renown*, and the elderly carriers *Furious* and *Argus* and other units of Force H, who were based at Gibraltar, rendezvoused with the *Victorious* and the convoy. After launching five Fulmars to the *Ark Royal*, the *Victorious* then spent two hours landing on and flying off Sea Hurricanes with RAF personnel who were transferring to the *Furious* and early next morning the units of Force H parted company. Group Captain O' Neill recalls the occasion: 'We were joined by the carriers *Argus* and *Furious* empty from an earlier delivery of fighters, together with the *Ark Royal* which was due to join us at Gibraltar and take off 260 Squadron. The four ships sailed together for a while and formed a picture which must have been quite rare in Royal Navy circles.'

On board the *Victorious* the Fleet Air Arm personnel then had the laborious task of taking all the various parts of the embarked Hurricanes from the hangar to the flight deck and by 6pm on Tuesday 10 June, 20 of them had been assembled and ranged there. Next morning, at 1.20am, the carrier passed Cape Spartel and at 5.25am she was secured alongside the main wharf at Gibraltar. For many members of the ship's company it was their first 'foreign' port and that afternoon leave was piped. However, there was still a great deal of work to be done, and with the *Victorious* berthed directly astern of the *Ark Royal* a large wooden ramp had been laid fore-and-aft between the flight decks of the two carriers so that 24 of the assembled Hurricanes could be wheeled across to the *Ark Royal*. Members of the ship's company who were able to get ashore found Gibraltar to be like another world with hot, sunny weather, well-stocked shops and seemingly unlimited supplies of food and beer in the restaurants and bars. Then, after dark, the streets and houses, shops and bars were all lit up, with no blackout. For many the bars and restaurants of Main Street were ideal, but others wished to explore this tiny colony further and Les Vancura recalls going ashore in the afternoon, catching a bus to the Spanish border and walking round the Rock to Catalan Bay where he went for a swim and had a slap-up meal.

During the early hours of Friday 13 June the air raid sirens at Gibraltar sounded and those on board the ships in the harbour had a rude awakening as they were called to Action Stations to engage a single enemy aircraft which dropped three bombs off the northern entrance to the harbour. Fortunately, despite what many may have considered an inauspicious day and date, there was no damage done or casualties suffered and at just after 11am the *Victorious* slipped her moorings and left Gibraltar. Half an hour later she was joined by the *Ark Royal*, the *Renown* and a destroyer screen, before the force set course north-east into the hostile waters of the Mediterranean.

Next morning, at between 10.45am and midday, in a position Lat 38° - 51'N/Long 03° - 54'E, approximately 70 miles south-south-east of Cape de Salinas, Majorca, *Victorious* flew off her 23 Hurricanes and the *Ark Royal* flew off the 24 aircraft which had been transferred three days earlier, escorted by a squadron of Hudsons which were

17

flying from Gibraltar en route to West Africa. Ahead of the RAF pilots involved in the operation lay a 700-mile flight to Malta, and they had been briefed that, in the event of a technical malfunction, they had only about a quarter of an hour in which to get back to their respective carriers for, as soon as the last aircraft was airborne, the carrier force would alter course 180° and steam at full power for Gibraltar. Forty-five of the Hurricanes made it to Malta, with one diverting to Algeria where the pilot was interned by the Vichy French, while the other was forced to ditch in the sea.

Victorious and the ships of Force H had an uneventful passage back to Gibraltar and she secured alongside the colony's detached mole at 12.35pm on Sunday 15 June. As soon as the brow was in place the crew of the German supply ship *Gonzenheim*, 11 officers and 52 ratings, were embarked for the voyage to the UK and a POW camp. At just after 7pm that evening the *Victorious* slipped her moorings and put to sea again, where she rendezvoused with the *Ark Royal* and *Renown* which had sailed an hour earlier. Over the next four days the force steamed out into the Atlantic where the Swordfish of 820 Squadron were transferred to the *Ark Royal* before, on the morning of Thursday 19 June, the two carriers parted company and the *Victorious* set course for Greenock where she anchored at 7.30am on Saturday 21 June. As soon as HM Customs had cleared the ship, leave was granted to a very grateful ship's company and Les Vancura recalls buying newspapers which carried reports of the *Bismarck* saga. HMS *Victorious* had certainly had an eventful start to her career.

The *Victorious* with Fulmars and Hurricanes on deck. HMS *Argus* is in the background. *(Fleet Air Arm Museum)*

Chapter Threee

Northern Waters

In the early hours of Sunday 22 June 1941, the day after the *Victorious* returned to Greenock, Germany launched the greatest attack in military history by invading the Soviet Union along a 1,800-mile front which stretched from the Arctic to the Black Sea. More than three million troops, 600,000 vehicles, 3,580 tanks and 1,830 aircraft were hurled against 4,500,000 Russian defenders who were caught totally unprepared. This time there was no elaborate deception as there had been for the invasion of Poland, but at 3am that morning when the guard was changed at the International Bridge at Brest-Litovsk, the German sentries gunned down their Russian counterparts instead of saluting them. Hitler was counting on scoring a quick victory and driving his massive armies to the Ural Mountains before the onset of winter, with the immediate objectives being Leningrad (St Petersburg), Moscow and Kiev. Despite heavy Soviet losses the Russian front line did not disintegrate as the Germans had expected, and on 1 July 1941 the German Army, in an alliance with Finland, launched two further military operations in the far north against the Arctic port of Murmansk and the Kandalaksha-Belomorsk railway line. The German offensive, launched from Kirkenes in northern Norway, quickly overran the port of Petsamo (Pechenga), which had been won for the Soviet Union in the Russo-Finnish War of 1939-40, and advanced into the Kola Peninsula. Suddenly Finland, for whom Britain and France had almost gone to war in late 1939, became part of the Axis and therefore a new enemy, and in London plans had to be drawn up to send military and medical aid to the northern Russian port of Murmansk where they were urgently needed.

Meanwhile, the *Victorious* had been enjoying a seven-day break at Greenock before leaving on Saturday 28 June to embark her squadrons. Upon her departure she set

The *Victorious* at sea with Fairey Albacores on deck. *(Fleet Air Arm Museum)*

course north to Scapa Flow and for most of the day she had to steam through thick mists with visibility so poor that speed was reduced to nine knots and at one point she was forced to stop engines when visibility dropped to zero. However, she eventually anchored safely at Scapa Flow during the early evening of the following day and it was the morning of Wednesday 2 July when the *Victorious* weighed anchor and put to sea again to land on the 12 Fulmars of 809 Squadron, and the 21 Albacores of 827 and 828 Squadrons. Despite the continuing poor visibility there was only one mishap when a Fulmar of 809 Squadron crashed on landing, but, fortunately, there were no casualties and by 11.15pm that evening she was anchored once more in the sheltered waters of Scapa Flow.

On Monday 7 July the *Victorious* started her work-up and flying training in earnest, and on most days during the following two weeks she was at sea carrying out gunnery exercises or flying trials. At just before 3pm on Saturday 12 July, during a day of flying exercises, a Fulmar crashed into the sea after its engine caught fire, and sadly both crew members, Sub-Lt Garner and Ldg Airman Powell, were lost. There was another, less serious accident, on Thursday 17 July when an Albacore of 827 Squadron crashed on landing, fortunately with no casualties this time. Five days later the carrier's second attempt to complete her work-up was cut short as operational needs took precedence.

Following the German attack on Russia came a request from the Soviet Government for a British naval attack on enemy troop transports in the northern Norwegian port of Kirkenes and at Petsamo. Both ports were of major strategic significance to the German forces involved in operations on the Finnish front, from the Kola Peninsula down to Leningrad. However, for the Royal Navy to undertake such a mission it would mean long voyages into the hostile waters of the Arctic Ocean where the ships would be within range of enemy land-based aircraft and where, during the summer months, they would not have the advantage of operating under cover of darkness. Although the politicians were keen to mount such an operation, Admiral Tovey was strongly opposed to sending his aircraft carriers, equipped as they were with obsolete aircraft, into enemy ports which were well defended with fast, modern fighters, but, as often happened during the Second World War, political considerations were put before sound military judgement and plans went ahead for the Fleet Air Arm attack on shipping in both ports. It was scheduled for Wednesday 30 July 1941.

The two aircraft carriers in the task force were the *Victorious* and the elderly, outdated *Furious,* and they were to be escorted by the cruisers *Devonshire* and *Suffolk,* together with a destroyer screen of HM Ships *Intrepid*, *Anthony*, *Antelope*, *Active* and *Achates*. The force, which was code-named Force P, left Scapa Flow at just before 11pm on Wednesday 23 July, bound for Iceland, from where the operation would be mounted. Next day the *Furious* scrambled two Fulmars in an effort to intercept two enemy aircraft, but with no success, and that afternoon the force encountered thick fog which, although it reduced visibility, kept the ships free from interference by enemy air reconnaissance. At 3am on the morning of Friday 25 July tragedy struck when, in thick fog off Iceland, the destroyer *Achates* hit a mine which virtually blew off her bows. Sixty-five members of her ship's company lost their lives, but she remained afloat and she managed to limp along by going astern before being taken in tow by the *Anthony*. The remainder of the force anchored off Reykjavik that afternoon and during the evening they steamed round to Seydisfjord (Seyoisjorour) on the west coast. Force P weighed anchor at 11pm on Saturday 26 July 1941 and set course for northern Norway, under the cover of thick fog, for the voyage which would take four days. The force reached the flying-off position, Lat 71° - 22'N/Long 33° - 02'E, approximately 80 miles north-east of Kirkenes, at about 1.30pm on Wednesday 30 July. Half an hour later 20 Albacores armed with either bombs or torpedoes, and escorted by 12 Fulmars, were launched to attack the shipping in Kirkenes Harbour, whilst on board the *Furious* nine Swordfish and nine Albacores, also escorted by Fulmars, were launched for Petsamo. Not only did the attack go ahead on a fine, clear day, but shortly before the start of flying operations it was known that Force P had been spotted by a Heinkel III reconnaissance aircraft. In spite of this intelligence political considerations took precedence over military judgement once again, and it was decided to go ahead with the raids because of their great importance.

During the briefing before the raids the aircrew had been told that there would be iron-ore vessels and, possibly, troop transports in the harbour, but for 828 Squadron it turned out to be far worse than they had expected, in fact it was a terrible ordeal. One Air Gunner recalls: 'The first time we realized things were turning nasty was when cannon shells started whizzing about our ears from the German gunners sitting above us on the crests of hills overlooking the fjords.' As the aircraft flew over the harbour it became apparent there were no big ships moored, but there were masses of German fighter-bombers which had been scrambled to intercept the Albacores and Fulmars from the *Victorious*. Frank Smith recalls: 'It was pure hell. The CO did a quick about-turn and we dumped our torpedo. We didn't care, there was nothing to fire them at anyway. As we turned, the third flight of the squadron got shot up almost immediately and out of the total of nine aircrew only one air gunner survived. In the second flight of three, six men were killed and only the three in the leader's plane escaped. We were just pounced on by the 110s and once they had finished with the others they chased our flight. The two wing aircraft broke to starboard

A bow view of HMS *Victorious* at anchor in Scapa Flow. She seems particularly popular with the seagulls. *(C. Minett)*

and headed for Russia. My impression at the time was that they both got shot down into the sea, but apparently one plane hit the sea, bounced back into the air and crash-landed in Russia.' Lt-Cdr (O) G. M. Haynes DSO RN also remembers the raid: 'We were about 80 miles from the coast and flew in, in fine formation, to find German 109Fs and 110s expecting us and flying around a conically shaped hill. Before we could get into the harbour proper, we were all under attack. A general mêlée occurred as we all forced our way into the inner harbour and dropped our fish. Some of us had already been downed and it was everyone for himself. We took off inland at first, then to the north to hug the high cliffs of the fjord, all the time under fire from a 109F. My pilot, Bobby Head, was concentrating on flying while I was the chap with the view and saying in the Gosport tube, "Slow down, Flaps down, Turn if you can." The shot was hitting the rock mostly ahead of us. Finally, we got to the head of the fjord and took departure by ourselves for the carrier *Victorious*. We thought at first that our opponent had taken off, but it was not to be. Instead of beam attacks he went up a couple of thousand feet and dived on us. Of course we were watching him and as soon as he was committed to the dive I would say to the pilot, "Down flaps, Turn right, Turn left," and from our flat-out speed of 100 knots we would decelerate to just above the stall, about 70 knots. The attacker would get his deflection all wrong and would have to go up again. After a few goes at us, he must have decided that other game was easier and disappeared, thank goodness. We had a few holes, but in an hour's time we were circling the carrier looking for our chums. Not one appeared. We could see them striking down the Fulmar fighters below and there were a couple of Albacores, but no great assembly of aircraft. We landed on and gradually the doom and gloom became evident - out of a squadron of 16 only three remained. We joined forces with the remains of the other squadron and after an attack on Tromso we returned to Scapa.'

With the air strikes from both the *Furious* and the *Victorious* having been set upon by waiting German fighters, heavy losses were suffered by the Fleet Air Arm, and the *Victorious* lost 11 Albacores and two Fulmars, whilst the *Furious* lost three Albacores. Nine of the aircrew from *Victorious* were killed in the attack, one of whom, Ldg Airman Griffin, died later of his injuries and was buried at sea that evening. Twenty-seven men were posted as prisoners of war and as a result of the action four officers were awarded the DSC, one petty officer was awarded the DSM, and an officer and air gunner were Mentioned in Dispatches. In the event, two German aircraft were shot down by the Fleet Air Arm aircraft, which was an achievement in itself given the superiority of the German aircraft.

The whole operation had been launched with inadequate aircraft and with no intelligence regarding the defences of the two ports, or even what shipping would be berthed there. Admiral Tovey's worst fears were realized and in his dispatch he wrote, 'The gallantry of the aircrews, who knew before leaving that their chance of surprise had gone and they were certain to face heavy odds, is beyond praise... I trust that the encouragement to the morale of

HMS *Victorious* at anchor in Seydisfjord. (C. Minett)

our allies was proportionately great.' However, Lt John Cooper RNVR, who was credited with the shooting down of one German aircraft and who was subsequently awarded the DSC, remembers that when Rear-Admiral W. F. Wake-Walker visited the wardroom of the *Victorious* to boost morale by pointing out that, despite the losses, the operation had shown the Russians that Britain was supporting them, the pilots and observers were so angry about the mishandling of the operation that they received what he had to say in stony silence. Lt Cooper went on to say, 'The atmosphere was both eerie and embarrassing and he had no course but to withdraw.' Perhaps the final comment on this sad saga should be to remember Ldg Airman Cyril Beer whose remains were found on the Russian-Norwegian frontier in 1988 and, with the ending of the Cold War, they were handed over by the Russian Government. So, 47 years after his death 'Jan' Beer was buried with full military honours in the Commonwealth War Graves Cemetery in Oslo.

At just after 1pm on Thursday 31 July, as Force P withdrew from the immediate area of operations off the Norwegian coast, an enemy Dornier 18 aircraft was sighted and successfully shot down by a Sea Hurricane from the *Furious*, which then transferred her serviceable aircraft to the *Victorious* and left for Scapa Flow. After embarking Rear-Admiral Wake-Walker from the *Devonshire* there was one further duty for the *Victorious*. At 1.10am on Monday 4 August, in a position Lat 71° - 37'N/Long 01° - 20'E, she launched three Fulmars armed with bombs to attack an enemy seaplane base at Tromso. On this occasion one aircraft did not return and the pilot, Lt Matthews, and observer, Sub-Lt Burrows, were posted as missing. By 6am that same morning the *Victorious* was steaming back to Seydisfjord, where she anchored at 6pm on Tuesday 5 August and Rear-Admiral Wake-Walker transferred his flag back to the *Devonshire*. Next day the *Victorious* weighed anchor once again, this time bound for Scapa Flow where she arrived in the early hours of Friday 8 August.

Following the disastrous raid on Kirkenes there was obviously a great deal of concern about the morale on board the *Victorious*, and during the morning of her first day at Scapa the C-in-C Home Fleet, Admiral Tovey,

This view shows Albacores on deck preparing for the attack on Kirkenes. *(S. Filmer)*

Fairey Fulmars ranging for take-off for the strike on Kirkenes. *(S. Filmer)*

visited the ship and took the salute at Divisions. Next day, Saturday 9 August 1941, at 1.35pm, hands were again piped to Divisions, but this time the Royal Marines detachment provided the Guard of Honour and, of course, the music. Just over half an hour later, at 2.15pm, His Majesty King George VI, who was visiting the Home Fleet at Scapa Flow, embarked and was escorted to the flight deck where the Royal Marines gave an impressive display of arms drill before the customary inspection. The King's visit to the *Victorious* lasted just an hour and he was obviously impressed by the turnout for, next day, as he left Scapa Flow in the destroyer *Inglefield*, the fleet were duly rewarded with the time-honoured custom of Splicing the Main Brace. However, it would soon be time to leave again

23

On 11 August 1941, His Majesty King George VI visited the *Victorious* at Scapa Flow, and after embarking...

...he inspected the Royal Marines detachment. *(S. Filmer)*

for with the first of the many Russian convoys due to depart Iceland for Archangel on 21 August, the *Victorious* was required at sea once again to assist with convoy escorts.

On Wednesday 13 August the ground crews of Albacore Squadrons 817 and 828 joined the ship, and next day at 9.40am *Victorious* weighed anchor and put to sea, escorted by the destroyers *Somali*, *Lively* and *Cassleton*, to embark the aircraft and carry out deck landing trials. For the next eight days the carrier sailed daily from Scapa Flow to carry out exercises off the north coast of Scotland. At 3pm on Saturday 23 August she left Scapa escorted by the cruisers *Devonshire* and *Suffolk* and the destroyers *Escapade* and *Inglefield*, to escort a convoy of 20 merchant ships which were carrying RAF Hurricanes, personnel, stores and equipment to Archangel. *Victorious* met the convoy during Sunday 24 August and three days later, in the Arctic

Ocean, the Albacores carried out dummy dive-bombing attacks and a Fulmar, which was patrolling over the convoy, crashed whilst attempting to land, but fortunately the aircrew were rescued by the *Inglefield*. At 8am on Saturday 30 August, with the convoy and the escorts in a position Lat 68° - 41'N/Long 41° - 31'E, at the mouth of the White Sea, the *Victorious* and the cruisers parted company and set course for Spitzbergen where, at just before 1am on Monday 1 September, they anchored in Sandam Bay. Les Vancura recalls that it was light until 1.30 am in the morning, and that during the day a number of the ship's company went ashore in the ship's boats but, as he records in his diary, 'No sign of life. Had to wade back to the boats, plenty of floating ice about.'

After just over 24 hours at anchor, at 1.54am on Tuesday 2 September the *Victorious*, together with her escorts *Devonshire*, *Suffolk*, *Inglefield* and *Impulsive*, weighed anchor to rendezvous with HM Ships *Argus* and *Shropshire* and three destroyers which were en route from Scapa Flow. The *Argus* had 24 Hurricanes embarked which were to be flown off to Archangel, but before the rendezvous took place the *Victorious* was diverted to attack enemy coastal shipping between Tromso and Hammerfest, although in the event she was then ordered to attack oil storage facilities at Hammerfest. The raid was to be carried out by 12 Albacores in two waves, with the first launch taking place at 12.34am on the morning of Wednesday 3 September 1941. However, 20 minutes later one Albacore crashed into the sea fully armed with its torpedo, but fortunately the aircrew were rescued safely. In view of the disasters which had occurred at Kirkenes and Petsamo the aircrew were briefed to attack only if there was cloud cover, and if they were satisfied that their approach had been undetected by the enemy. In the event, with clear skies over the coast and having been challenged by enemy anti-aircraft defences, both strike forces returned to the *Victorious*, for with strong fighter defences in the area it would have been suicidal and pointless to continue and by 2.39am all eleven Albacores had safely landed on. However, at 5.32am enemy aircraft were seen to be shadowing the *Victorious* and three Fulmars were launched in an attempt to intercept them. Throughout the day, as the *Victorious* steamed north towards Bear Island, enemy aircraft continued to shadow her and at 2.50pm one Fulmar which had been deployed to intercept them crashed into the sea while attempting to land on, but happily the aircrew were rescued uninjured, although very cold and wet. The raid achieved nothing, apart from alerting the enemy defences in northern Norway to the carrier's presence in the area, although a Dornier 18 flying boat had been shot down during the morning. By 8pm that evening the *Victorious* was in a position Lat 73° - 56'N/Long 12° - 06'E, south-east of Bear Island, and during the morning of Friday 5 September she rendezvoused with the elderly carrier *Argus*, the cruiser *Devonshire* and three destroyers, after which the whole force steamed towards Murmansk. The *Argus* was carrying 24 Hurricanes which were assembled on board and in just over two hours, between 5.20am and 7.30am on Sunday 7 September, they were flown off in four waves of six aircraft to Murmansk. Once the aircraft were away the *Argus* and her escorts set course for Seydisfjord, and *Victorious*, *Devonshire*, *Suffolk* and the destroyers steamed for Spitzbergen, where they anchored during the evening of Tuesday 9 September. Les Vancura recalls that before she left, the *Argus* transferred two Grumman Martlets to *Victorious* in exchange for two Fulmars, as the American fighters were faster and would be better able to intercept the shadowing enemy aircraft. He also has vivid memories of the sun setting on the snow-capped mountains of Spitzbergen, a beautiful and awe-inspiring sight. Next morning, at 9.35am, Rear-Admiral Wake-Walker transferred from *Devonshire* and hoisted his flag in the *Victorious* then, after refuelling from an RFA, the force weighed anchor and set course, once again, for the coast of northern Norway.

This time the *Victorious*' squadrons had been detailed to attack shipping in Vestfjorden and the port of Bodo with bombs and torpedoes. The flying operations commenced at 1am on the morning of Friday 12 September, when the carrier was in a position approximately 40 miles west of the Lofoten Islands. The strike force of 12 Albacores from 817 and 832 Squadrons sank two coastal vessels, one of which was the Norwegian passenger steamer *Boro*, with a large number of casualties. Five of the Albacores flew further south to the small town of Glomfjord where they attacked a power station and an aluminium works, which was set on fire. This time the raids were a success and at 3.45am all 12 aircraft landed on safely having encountered no serious opposition. As the raid was so successful Admiral Wake-Walker had intended to send in a second strike, but at 8am that morning a Heinkel III was detected and the two Martlets which had been transferred from the *Argus* were launched to intercept it. After a long chase Lt J. W. Sleigh managed to shoot down the shadower, but not before the enemy defences in the area had been alerted. Unfortunately, when landing on at 9.30am, one of the Martlets crashed, luckily with no casualties, and with more enemy aircraft starting to shadow the *Victorious* it was decided to cut short the operation, and the carrier left the coast of the Lofoten Islands and set course for Scapa Flow, where she anchored at just after 7pm the next day, Saturday 13 September.

The *Victorious* remained at her anchorage for nine days, during which time she received a visit from the C-in-C Home Fleet and embarked stores and provisions, which Les Vancura noticed had come from the still neutral USA. Then suddenly, during the early afternoon of Tuesday 23 September, with only a few hours' notice, the carrier

An Albacore being launched from the *Victorious*. (S. Filmer)

weighed anchor and proceeded to embark the Albacores of 817 and 832 Squadrons. In company with the battleships *King George V* and *Malaya* and the destroyers *Punjabi*, *Eskimo*, *Aurora*, *Somali* and *Matabele*, she then set course for Hvalfjord, Iceland, where the whole force anchored during the afternoon of Thursday 25 September. Also at anchor were units of the US Navy, including the battleship *Idaho*, which was clear evidence of how the 'neutral' USA was carrying out vital convoy escort duties. Two days later the aircraft carrier USS *Wasp* entered harbour and during their stay exchange visits were arranged for the officers and men of both aircraft carriers. Les Vancura went on board the *Wasp* and was immediately impressed by the spotlessly clean ship which he described as, 'all chrome and pale blue paint.' He recalls that all the aircraft had been ranged on the flight deck so that the hangars could be turned into recreation spaces and basketball and badminton tournaments were in full swing. With their customary generosity the US Navy also treated their guests to a film show and a slap-up six-course supper, the like of which they had not seen for a long, long time. Unfortunately, Les Vancura did not get an opportunity to make a visit to the *Idaho*, for at 8am on Thursday 2 October the *Victorious* weighed anchor for two days of flying exercises off Iceland, before setting course for Norway during Saturday 4 October, in company with *King George V* and the 6th Destroyer Flotilla. Flying exercises were carried out during the passage and when, sadly, one of her Albacores failed to return on the evening of Sunday 5 October, the crew, Sub-Lt P. G. P. Dundas RN and Ldg Airman M. S. Eastment, were posted as missing. By the morning of Wednesday 8 October the *Victorious* was once again 40 miles west of the Lofoten Islands and, in very rough weather, a strike force of Albacores was launched to attack shipping in Vestfjorden. In the inclement weather only three aircraft were able to press home their attack on a coaster, but a second strike force of eight Albacores which took off at 11am attacked two merchantmen being escorted by anti-aircraft ships, and damaged both of them. Once again the operation had been successful and the *Victorious* then set course for Scapa Flow, where she anchored safely during the morning of Friday 10 October. Two days later, during the afternoon of Sunday 12 October, she weighed anchor once again and this time set course for Greenock. Three hours after leaving, whilst she was in the Pentland Firth, night-flying exercises were commenced and at 9.20pm an Albacore crashed into the sea about a mile off the carrier's port beam. The *Victorious* remained in the vicinity of the accident while the escorting destroyers searched the area, but they were unable to find any trace of the aircrew, Lt J. E. R. Flowers RNVR and Ldg Airman W. James, and at 10pm the search was abandoned. The following day the carrier anchored off Greenock where most of the ship's company were able to get ashore and some of the lucky ones even managed overnight leave in Glasgow. However, six days later and accompanied by the destroyers *Lightning* and *Laforey*, the *Victorious* set course for Scapa Flow again. Soon after leaving the Clyde she ran into severe south-westerly gales with wind speeds of over 56

An unusual view of the *Victorious* with Fulmars on deck and an Albacore overhead. (*Fleet Air Arm Museum*)

knots and enormous waves capped with dense white foam. As the storm raged the sheer severity of the seas stove in five scuttles, carried away one seaboat and smashed five more. One of the mobile cranes on the flight deck broke its chains and was washed over the side, carrying with it an ammunition locker. So severe was the storm that it even damaged the carrier's superstructure and everyone was very relieved to reach Scapa once again during the early evening of Monday 20 October.

Some of the ship's company were lucky enough to have leave to look forward to, but the majority could only enjoy a short respite before the *Victorious* weighed anchor on Thursday 23 October to carry out flying exercises in the Pentland Firth. At 10.15am that morning an Albacore which was on a navigating exercise crashed into the sea with the loss of her aircrew, Sub-Lts P. D. Say and K. Hopkins, and Ldg Airman N. Weldon. Two days later the carrier put into Scapa Flow and on Sunday 26 October, during a visit to the ship by the C-in-C Home Fleet, the serviceable aircraft were catapulted off to RNAS Hatston while the ship was at anchor. On Saturday 1 November the Board of Inquiry was held into the loss of the Albacore nine days previously, and concern was expressed about the number of fatal accidents which had occurred as a result of engine failures in the Albacores. Two days later the *Victorious* sailed again, in company with the battleship *King George V* and the cruisers *Sheffield* and *Edinburgh*, bound for Hvalfjord in Iceland. Once at their destinations they were joined by units of the US Navy, including the battleships *Idaho* and *Mississippi*, which made a fleet of 27 warships. During the whole of November the *Victorious* operated from Hvalfjord, which kept her and other heavy units close at hand should the *Tirpitz* attempt a breakout into the Atlantic as her sister ship *Bismarck* had done just six months earlier. On Monday 17 November the *Victorious* left her anchorage in Hvalfjord at 7.30am to carry out flying and gunnery exercises. She was in company with the battleship *King George V* and at 12.50pm that day 22 Albacores were launched to carry out dummy torpedo attacks on the battleship. By 2.50pm, with the exercise completed, the aircraft were being landed back on again, but as Albacore 3R made her approach the aircraft suffered an engine failure and by skilful handling the pilot was able to make a forced landing very close to the *Victorious*.

An aerial view of HMS *Victorious* off Spitzbergen in September 1941, while supporting Russian convoy PQ1.

(C. Minett)

Almost immediately the destroyer *Impulsive*, which was acting as a planeguard, was on the scene and all three crew members were rescued safely. In the evening, after the *Victorious* had returned to Hvalfjord, they were taken back on board the carrier. By the last day of November the *Victorious* had returned once again to Scapa Flow and more members of the ship's company were able to take well-earned leave, while the *Victorious* undertook some daily outings to the Pentland Firth, mainly to carry out anti-aircraft gunnery exercises.

However, by this time the relative calm of the Pacific region and South-East Asia had been shattered by the devastating Japanese attack on the US Naval Base at Pearl Harbor and simultaneous invasions of Malaya and the Philippines, which changed the course of the whole war. Although these momentous events in the Far East were a long way from Scapa Flow and the day-to-day life of the men on board HMS *Victorious*, with the German declaration of war against the United States the whole might of the US Navy's Atlantic Fleet was now actively engaged on the Allied side. During January 1942 the routine for the *Victorious* changed very little, with daily exercises in the Pentland Firth. During one of these, at midday on Saturday 17 January 1942, whilst landing on the Albacores of 809 and 817 Squadrons, the propeller of one aircraft shattered and a young air mechanic was fatally injured when he was hit by large fragments of shrapnel.

That same afternoon the carrier anchored in Scapa Flow and the body of Air Mechanic C. W. Carr was taken ashore, together with the damaged aircraft, after which the *Victorious* set sail again, in company with the *Rodney*, *King George V*, *Nigeria*, *Kenya*, *Sheffield* and a destroyer screen, to patrol the Denmark Strait, for the spectre of the *Tirpitz* loose on the Atlantic was tying down all the major units of the Home Fleet.

Most of February was spent carrying out patrols from Hvalfjord and at just after 5am on Thursday 19 February, after a delay of two hours whilst the battleship *King George V* sorted out a fouled anchor, the *Victorious* left the Icelandic anchorage to escort a Russian convoy and then to carry out an offensive sweep against shipping in the Tromso area. Admiral Tovey was flying his flag in the battleship *King George V*, and also in company was the cruiser *Berwick* and a destroyer screen. Just two days out of Hvalfjord signals were received to the effect that the pocket battleship *Admiral Scheer*, and the heavy cruiser *Prinz Eugen* were at sea in the Skagerrak and about to head north to join the *Tirpitz* at Trondheim. However, in the event, the two German ships anchored in a fjord near Stavanger and the *Victorious*, escorted by the *Berwick* and four destroyers, *Bedouin*, *Ashanti*, *Eskimo* and *Punjabi*, made her way to a flying-off position of Lat 63° - 40'N/Long 02° - 45'E, about 100 miles off the headland of Stadtlandet. During her approach the barometer was falling and the weather

28

Here the *Victorious* is seen ploughing through heavy seas in northern waters, autumn 1941.
(*C. Minett*)

continued to deteriorate with a rising wind and frequent snow squalls which at times reduced visibility to just two hundred yards. At 12.45am on Monday 23 February, having reached the designated position and with the first range ready for take-off, Captain Bovell had to postpone the first launches because of a particularly heavy snow squall which obscured the horizon. He sent for the meteorological officer who, although he had no fresh information, considered that the weather would be much better off the Norwegian coast. At 1am, with the snowstorm having abated, Captain Bovell ordered the first range of ten Albacores of 832 Squadron, armed with torpedoes and under the command of Lt-Cdr A. J. P. Plugge RN, to be launched. This in itself was no mean feat with the ship pitching into 31ft waves but, after experiencing great difficulties in getting signals through and turning the force back into the wind, at 1.45am the second range of seven Albacores of 817 Squadron, also armed with torpedoes and under the command of Lt-Cdr D. Sanderson DSC RN, were launched. Once the second range had been flown off, the *Victorious* and her escorts increased speed to 28 knots and rejoined the *King George V* off the Shetlands just over eight hours later.

The orders which had been given to 832 Squadron were to make for the Norwegian coast at Stadtlandet, and to search for the enemy ships as far south as their endurance would permit and then to land at an RAF airfield at Sumburgh in the Shetland Islands. During the search they were to stay roughly ten miles off the Norwegian coast, and to keep as low as possible to avoid enemy radar. They were also ordered to attack with torpedoes any enemy warships which they encountered. The orders given to 817 Squadron were identical, except that they were to keep 20 to 30 miles off the coast on their sweep south. All aircraft were ordered to remain in formation, but if they became separated they were to carry out the operation independently. In fact, this is exactly what happened to Albacore 4C, piloted by Sub-Lt James D. Landless RNVR, and crewed by Sub-Lt David Harvey RNVR and Ldg Airman Thomas Armstrong, which was the last aircraft of 832 squadron to take off, straight into an approaching snowstorm. When he was clear of the squall the rest of the squadron were nowhere to be seen and, as he circled the ship, Captain Bovell fully expected him to land back on again. However, the aircraft set off on the mission and completed the whole of the arduous night sweep alone, landing at Sumburgh four and a half hours later.

Although the weather was reasonably good when the rest of the aircraft set out, it soon deteriorated and the coastline disappeared beneath clouds and mist. Unfortunately, conditions worsened even further when heavy snowstorms were encountered, and at 1.58am the ASV aircraft, 4H, picked up three echoes, with a clear outline, at a range of eight miles and some 18 miles from the coast. They were presumed to be ships and the formation ahead altered course to pass over them but, owing to poor visibility, nothing was sighted. At 2.45am, the formation of seven aircraft reached a position approximately 20 miles west of Utvoer Light, which was the limit of their endurance and, having sighted no enemy shipping, they turned and set courses for the Shetlands. However, fifteen minutes later a light was sighted about 12

HMS *Victorious* in Arctic waters during the winter of 1941/42 showing her ice-encrusted gun turret and guard rails.
(C. Minett)

Following her attack on the *Tirpitz* on 9 March 1942, the *Victorious* herself was the target of German aircraft, one of which managed to get through the defensive screen and drop two bombs which only just missed the carrier.
(C. Minett)

miles away and the formation turned to investigate but found nothing. (It was later established that the light was a flare from 4C which was operating independently). The flight to Sumburgh was extremely difficult and at 4.15am Albacore 4A, Lt-Cdr Arthur J. P. Plugge RN, dropped a parachute flare, before diving towards the water and crashing into the sea. Fortunately, the remaining six aircraft landed safely at Sumburgh.

The aircrew of 817 Squadron faced an equally daunting task and, in their case, it was marred by tragedy. The squadron had taken off at 1.46am in reasonable conditions, albeit showery, and generally visibility was good. However, once off the Norwegian coast the weather deteriorated rapidly and they too found the coast obscured by cloud, so they altered course to starboard. Then at 2.50am, in the midst of a heavy snowstorm, an explosion was heard and a flash was seen which was believed to have been caused by aircraft 5F and 5H colliding, for neither of them returned from the operation. Once in the vicinity of Sando Light the aircraft set course to return to Sumburgh and as 5K broke out of the cloud cover it encountered a small naval craft at which it released its torpedo, although without result. While returning to the Shetland Islands thick cloud and heavy snow was encountered, but the five remaining Albacores landed safely at the Shetlands base.

In his report to the Admiralty Captain Bovell was critical of the quantity and quality of intelligence he received and the poor weather forecasting prior to the mission, declaring, 'My views on the prospects of the success of this operation are given in my signal.... These prospects, which were slender at best, were rendered hopeless by the weather, and, had I known the conditions over the Norwegian coast, I would never have flown off the striking force. Long odds must be taken in war and losses accepted, but I submit that the service cannot afford to squander its slender capital of trained Fleet Air Arm personnel when information is available that there is no chance of success.' Regarding intelligence he stated, 'One of the chief lessons of this war has been that, if an air attack is not to be a costly "blow in the air", recent and accurate reconnaissance is essential. In no operation (except that which resulted in the sinking of the *Bismarck*) do I feel that I have received the quantity or quality of information necessary for success. On this occasion RAF aircraft had been over the Norwegian coast at frequent intervals for the past few days; Coastal Command had had aircraft operating from the south as far north as Stadtlandet up to midnight 22nd/23rd. Yet no report on the weather off the Norwegian coast had been included in the routine synoptic reports since 05.00 21st, and no special report was sent to me (though these must have been available).' However, in his remarks on his flying crews he had this to say: 'The courage and determination displayed by all flying crews carrying out the full sweep in the face of such conditions deserve the highest praise. Although the sweep failed in its object of locating and attacking enemy ships, and unfortunately resulted in the loss of three aircraft and flying crews, it served a purpose in giving confidence to the remainder who took part, in their own ability and in their aircraft.'

At 11am on Wednesday 4 March 1942 the *Victorious* left Scapa Flow again in company with the battleships *King George V*, *Duke of York* and the *Renown*, the cruiser *Berwick* and a screen of nine destroyers. This powerful force was to provide cover for a convoy of 15 merchant ships, PQ12, which had sailed from Reykjavik three days earlier bound for Murmansk, and convoy PQ8 which had left Murmansk for the UK on the same day. The sheer size of this covering force gives some idea of just how much disruption was caused by the presence of the *Tirpitz* at Trondheim, with the ever-present possibility that she would break out into the Atlantic or operate against the Arctic convoys. During the winter of 1941/42 the convoys to and from northern Russia had operated with little interference from enemy forces, but they had to contend with some of the worst natural conditions to be found anywhere in the world. Between Greenland and Norway the gales, rain, sleet and snowstorms sweep north-east creating seas with enormous waves. Fog is frequent and the temperature rarely exceeds 2°C (35°F), and in fact temperatures are often so low that the sea spray freezes immediately when it strikes a ship's superstructure. From the polar regions pack ice sweeps down and creates problems in the approaches to Archangel, where ships cannot proceed without the assistance of an ice-breaker. During the war, in order to avoid enemy aircraft, it was necessary to route convoys to the north of latitude 77 where, for at least 115 days of the year, there is perpetual darkness. Indeed, during *Victorious*' operations in February 1942 the temperatures on the flight deck had registered 20 to 45 degrees of frost, and everything had become encrusted with tons of ice.

It was into such conditions that the covering force sailed in early March 1942, and both their presence and that of the convoys was known to the enemy which had sent reconnaissance aircraft to shadow the Allied vessels. During the evening of Friday 6 March a report was received that a German heavy ship was at sea north of Trondheim, and next day it was confirmed that the *Tirpitz* had left the Norwegian port. In his diary Les Vancura noted, 'Action Stations, an enemy battleship in the vicinity.' At noon that day the two convoys were 200 miles south-west of Bear Island and the *Tirpitz* was, unknown to both sides, only 70 miles away from them. All that day, and during the next day, the search for the German battleship and *Tirpitz*'s search for the Allied convoys continued. However, in the early hours of Monday 9 March a signal was received in *Victorious* from the C-in-C Home Fleet which read: 'Expect *Tirpitz* in position Lat 68° - 15'N/Long 10°- 38'E

The vast expanse of *Victorious'* flight deck is covered in snow. The battleship HMS *Duke of York* is in the background. *(C. Minett)*

Footprints in the snow. The flight deck, winter 1941/42. *(C. Minett)*

at 07.00 A/9 steering southwards. Report proposals.' In fact, Admiral Ciliax in *Tirpitz* had decided to return to Trondheim and Admiral Tovey's information had been received as a result of an 'Ultra' intercept. Captain Bovell's reply to the C-in-C was: 'Propose fly off searching force of six aircraft at 06.30 to depth of 150 miles sector 105 to 155 degrees. Fly off striking force of 12 as soon as ranged about 07.30. Ship maintain present course and speed after searching force flies off till 10.00 then turn 315 degrees 26 knots. Fighters remain with ship. Consider 12 aircraft maximum that can be ranged in present weather.' Admiral Tovey was obviously well pleased with this plan for he replied: 'A wonderful chance which may achieve most valuable results. God be with you.'

At 6.45am on 9 March six Albacores were flown off to commence the search for the *Tirpitz*, and they were followed at 7.35am by 12 Albacores armed with torpedoes. At 8am search aircraft 'Duty F' sighted the *Tirpitz* and made the first report, and from then on the German battleship was constantly shadowed and reported. At 8.42am the striking force sighted the *Tirpitz*, which was accompanied by a destroyer, and they closed to attack. The approach was made upwind, and as the *Tirpitz* was actually steaming upwind the Albacores only closed her at some 30 knots. After climbing up through the clouds, the strike force came back out of the cloud cover astern of the battleship and both the leading sub-flights attacked the port side. This enabled the *Tirpitz* to turn to port under full wheel and to comb the torpedo tracks, thus forcing the two rear sub-flights, which were coming in on the starboard side, to make a long detour into the wind to get into a favourable position for dropping their torpedoes. This delayed their attack and it gave the *Tirpitz* time to reverse her wheel and comb the second lot of torpedoes. Unfortunately, none of the torpedoes hit the target and two Albacores were lost in the attack, having been shot down by close-range anti-aircraft fire. One of them was sighted in the water after the attack, but the other was not seen again after the dive to attack. Sadly one of the crew members of the former aircraft was spotted by members of the *Tirpitz's* ship's company, sitting helplessly on the top wing of his aircraft, but having escaped the attack there was no question of the *Tirpitz* stopping to pick him up and in the freezing cold waters he could not have survived for long. During the attack on the *Tirpitz* the German battleship's own Arado seaplane attacked the search Albacore 'Duty F', without causing any damage, and by 10.45 am all the surviving Albacores had landed safely back on the *Victorious*, which then set course for Scapa Flow. However, less than an hour later a hostile aircraft was sighted shadowing the carrier and, although three Fulmars drove it off, at 3.45pm three JU 88s attacked the *Victorious*, one of which managed to get through the defensive screen and drop two bombs which only just missed the carrier. The remainder of the voyage back to Scapa passed without incident and the *Victorious*, escorted by five Beaufighters, anchored safely in Scapa Flow at just after midnight on the morning of Friday 11 March.

Over the next ten days the *Victorious* put to sea for flying exercises on three separate days, and on the morning of Sunday 22 March she left the Flow once more, this time with the battleships *King George V* and *Duke of York* and the battlecruiser *Renown*, to cover the passage of convoy PQ13. As usual on such operations anti-submarine patrols were constantly flown and at 7.10am on Tuesday 24 March whilst flying off such a patrol, an Albacore, armed with depth charges, caught the carrier's island with his wing tip and crashed into the sea, where the depth charges exploded with such violence that the explosion was felt throughout the *Victorious*. None of the aircrew survived, but the body of one of them, Sub-Lt D. S. Davies RNVR, was recovered by the destroyer *Icarus*. Next day, when the *Victorious* was off Iceland, the weather began to deteriorate rapidly and conditions on board are described by Ray Barker, who was on the captain's secretarial staff: 'Fierce winds and high seas whipped at a fine angle across the front of the fleet. The destroyers were the first to feel the effects and began pitching, tossing and rolling to an alarming extent. Waves lifted their bows clear of the water for almost a third of their lengths and, as they plunged down, their entire forecastles became submerged and huge cascades of freezing water rained down on their open bridges. Speed was reduced because they could no longer make normal progress through the mountainous seas. The gale worsened and the sea became a raging tempest, frightening in its ferocity. Ahead of us the *King George V* was rolling, twisting, wallowing like a hippo in a mud bath. Occasionally she would disappear and we would watch for her emerging, shaking herself free of water from another gigantic trough. As it reached 100 miles an hour the wind began to bring real danger. Our bows, 40 feet high, began to dip into cavernous swells and enormous waves, rising up to 70 feet, thundered over the forward end of the flight deck. The pressure on the bow plates was intense and each time the ship shuddered, her 30,000 tons almost stopped in its tracks; we wondered what damage we would eventually suffer. Already in the bow sections the steel plates were buckling and shipwrights were busy shoring them up with timber. Their working conditions were appalling as the thunder of the massive waves lifting the bows reverberated around their working spaces and the rise and fall of 60 to 70 feet produced nausea and disorientation. Water was seeping through as rivets sprang and eventually when the timber props were in place, the watertight screen doors and hatches were clamped closed to isolate the bow section.

Aircraft were ranged on the flight deck armed with torpedoes, and flight deck crews were at full strength on deck, ensuring the aircraft remained secure in the atrocious conditions. These men hung onto chocks and securing lines whilst being virtually torn from the aircraft. At the forward end of the flight deck we were equipped with wind shields which were hydraulically raised and, when in position, were intended to deflect the wind and protect parked aircraft. Suddenly, without warning, these massive steel shields were uprooted, lifted clear of the deck and became airborne. It was an amazing but terrifying sight and had they been swept back at deck level the effect could have been catastrophic. In the event they landed near to the *Duke of York*, astern of us. Standing on the bridge was alarming. At that height those of us up there rolled through the widest arc of anyone on board and it was dizzying. We were regularly listing either to port or starboard at up to 15 degrees and it was almost impossible to stand. Then came an enormous list of 21 degrees and I felt we would just carry on over and capsize. The hairs stood up on the back of my neck, my heart seemed to pump to bursting point and my stomach turned to water. Would the roll ever end?

Slowly the list began to reverse and the ship returned to its proper attitude and down below there was pandemonium as lightly secured objects were thrown about. The galley staff were unable to man the cooking ranges and we had no hot food.' Fortunately, by the next day the storm had abated, but after running through snowstorms and freezing temperatures the *Victorious* was encrusted in tons of ice.

By the morning of Saturday 28 March the *Victorious* had returned to Scapa Flow, but having sustained damage to her bow plating and bulkheads during the storm she required dockyard attention and that same evening she sailed for Rosyth, arriving alongside the South Wall in the Naval Dockyard at just before 1pm the following day.

It was 12 months to the day since she had first commissioned at Newcastle upon Tyne and it had been an extremely busy time for all the ship's company and particularly for the aircrews who, from the very first weeks of the ship's career, had been required to take part in the most difficult and dangerous operations.

Snow clearance on the flight deck. *(C Minett)*

Chapter Four

'Operation Pedestal'

Although the *Victorious*' stay at Rosyth Dockyard was limited to nine days, it enabled most of the ship's company to see something of the city of Edinburgh which was easily accessible by ferry. All too soon, however, during the evening of Tuesday 7 April, the *Victorious* was under way again making an overnight passage north to Scapa Flow where she arrived the next morning. By now the US Navy's Atlantic Fleet was much in evidence and Les Vancura recalls going ashore to watch two American teams playing baseball, and an American team visiting the *Victorious* to play deck hockey. Ray Barker remembers his surprise at finding out that the US Navy ratings were, '...paid about three times as much as us, secondly the materials of their uniforms, shirts, shoes and coats was vastly superior, and lastly they seemed more relaxed - the relationship between ratings and officers was less formal than ours.' It would not be long before the ship's company from *Victorious* would themselves be forming part of the US Navy's Pacific Fleet.

On 9 April the *Victorious* put to sea for the day to embark her aircraft and to carry out deck landing practice, during which a Fulmar crashed into the sea, but the pilot was safely rescued by the destroyer *Eskimo*. Three days later, on Sunday 12 April, she sailed once again with the battleships *Duke of York* and *King George V* to cover the Russian convoys. During the four days of this patrol four aircraft were damaged in crashes whilst landing, but fortunately no personnel were injured. However, on Monday 20 April, while exercising in the Pentland Firth, a Fulmar crashed on landing with the loss of its pilot whose body was recovered by the detroyer USS *Plunkett*. The *Victorious* sailed from Scapa Flow, once again to cover the Rusian convoys, on Tuesday 28 April and this time she was accompanied by the battleship *King George V*, the cruiser *Kenya* and the US Navy's battleship *Washington* and the cruiser *Tuscaloosa*, as well as an assortment of British and US destroyers which included the Tribal-class vessel HMS *Punjabi*. At 3.51pm on the afternoon of Friday 1 May, when the force was in a position Lat 66° - 33'N/Long 10° - 41'W and steaming in thick fog, there was a submarine alert and the *Punjabi* cut across the bows of the *King George V* in order to engage the contact with depth charges. However, the destroyer's captain clearly misjudged the speed of the battleship and *King George V* collided with the *Punjabi*, cutting her in half. The *Victorious*, steaming astern of the battleship, had to alter course sharply in order to avoid survivors from the destroyer who were desperately swimming in the icy waters of the Norwegian Sea. Ray Barker remembers that, 'The fog was thick, swirling and cold. It had a freezing clamminess that left beads of ice-cold vapour on eyebrows and eyelashes. I looked down at the water - almost carbon black, terrifyingly evil. There were men drifting by, and carley floats were cut from our bridge superstructure and individual seamen threw their lifebelts into the water. From their cries we knew the men were British, but in no time they were astern of us and out of sight in the thick dark grey fog.' Thirty-eight men lost their lives in the tragedy, but it could have been far worse, considering how quickly the destroyer sank. Four days later the *Victorious* anchored in Scapa Flow where Admiral Tovey transferred his flag from the *King George V* to the *Duke of York* as the former had to go into dry dock for repairs to a massive 20ft gash in her stem.

On Monday 11 April a number of Admiralty officials were embarked in the carrier and next morning the *Victorious* sailed into the Pentland Firth to carry out deck landing trials with the new Supermarine Seafire. Following the success of the Hawker Sea Hurricane in carrier operations, it was decided to develop a naval version of the Spitfire. In late 1941 a Spitfire was fitted with an arrester hook beneath the fuselage and flown successfully onto HMS *Illustrious*. While the Spitfire's narrow track landing gear made it more difficult to operate aboard carriers than the Hurricane, the concept appeared sound and work began at Hamble near Southampton to convert a number of RAF Spitfires. Other aircraft were modified by Supermarine on the production line and they were designated as Seafire IBs. With a maximum speed of 352mph over the Fulmar's 247mph, and with two 20mm cannon in addition to four ·303 machine-guns, plus provision for 500 pounds of bombs, the Seafire would be a formidable weapon for the Fleet Air Arm. The landing trials on the *Victorious* were completed successfully and by 8pm the same evening the carrier had anchored in Scapa Flow where the Admiralty trials party disembarked. However, once in service the Seafire proved to be an unsatisfactory aircraft in many ways, suffering an unusually large number of deck landing accidents. One problem was the weak undercarriage, and many propellers were shattered when the planes' noses dipped down towards the deck as the hooks picked up the arrester wires. The Seafire's long nose also obstructed a pilot's view when landing.

In the early hours of Friday 15 May the *Victorious* received orders to put to sea and at just after 3.30am, together with the battleship *Duke of York*, she left the anchorage at Scapa Flow bound for the Barents Sea. On

35

Fairey Albacores on deck. (C. Minett)

USS *Washington*, and next day, after landing on another Seafire, she anchored in Hvalfjord. Ray Barker recalls the Seafire landing on: 'I watched fascinated as the RAF Wing Commander approached us. Because of the mid-plane placement of the cockpit it was very difficult for the pilot to see our flight deck on the long run in from astern, and he had to fly the aircraft crab-wise and then straighten up at the last moment. All of us prayed for a successful trial, but we need not have worried as the Seafire landed safely, picked up the second arrester wire and was brought to a halt. The flight deck crew cheered and clapped. Then the take-off, a circuit, landing again perfectly. Finally the Seafire was attached to our port side catapult and launched off the deck at speed. It dropped immediately it cleared the bow but as we anxiously scanned the sea ahead of the ship, we saw the aircraft flying, some 20 feet above the waves and almost creating a wake.' It was Wednesday 20 May before the *Victorious* put to sea again for gunnery exercises, but the day was marred by the loss of an Albacore which crashed into the sea off the ship's starboard beam. Unfortunately, all the aircrew were lost. Later that afternoon Les Vancura remembers the outward bound convoy of at least 60 merchant ships steaming past the *Victorious* in line ahead and all of them dipping their ensigns and crews cheering as they slowly passed by. The whole procession of ships took two and a half hours to complete their steam past, following which the *Victorious* returned to her anchorage at Hvalfjord. During her stay Les Vancura recalls that Lieutenant Douglas Fairbanks Jnr USN visited the carrier. He was the pilot of one of the USS *Washington's* three aircraft and Les remembered him looking every inch the Hollywood movie star as he passed by on the way to the wardroom, with his peaked cap perched jauntily on the back of his head.

the previous day German torpedo bombers had attacked the cruiser *Trinidad* west of Bear Island as she was returning to Scapa from Murmansk. She had already been torpedoed whilst escorting convoy PQ13, but she had managed to make Murmansk where temporary repairs could be carried out. This second attack had virtually disabled her and before the *Victorious* was able to get anywhere near the scene the crippled cruiser was sunk by British gunfire and torpedoes. On Saturday 16 May the *Victorious* joined the US battlefleet which included the

The first two weeks of June saw the *Victorious* operating out of Hvalfjord and by the second half of the month she was back at Scapa Flow and carrying out deck landing practice and training with her Fulmars and Albacores, and also a Sea Hurricane of 885 Squadron. On Monday 29

The *Victorious*, with HMS *King George V* astern, steams through heavy seas in northern waters, April 1942. (C. Minett)

June Admiral Sir Bruce Fraser, Second in Command, Home Fleet, hoisted his flag in the *Victorious* and the rest of the Hurricanes of 885 Squadron were embarked. Two days earlier the ill-fated convoy PQ17 had sailed from Hvalfjord bound for Russia and, once again, the *Victorious* and heavy units of the Home Fleet and the US Atlantic Fleet were detailed to provide distant cover for the merchantmen. The convoy consisted of 22 American, eight British, two Russian and two Panamanian ships and one Dutch merchant vessel. As well as a close escort of four cruisers and a number of destroyers, there were three small convoy rescue ships which had been specially fitted for picking up the crews of torpedoed merchant vessels. This considerable, and extremely valuable, convoy had been spotted by a German submarine and by enemy aircraft on 30 June and three days later the Luftwaffe began the first phase of air attacks on the convoy. At the same time Grand Admiral Raeder successfully requested a sortie using ships of his Northern Fleet. Code-named 'Operation Rösselsprung' it involved the *Tirpitz*, *Admiral Hipper*, *Admiral Scheer* and *Lutzow* and the destroyers *Hans Lody*, *Friedrich Ihn*, *Theodor Riedel* and *Karl Galster*. This powerful force was to be 100 miles north of North Cape as the convoy passed 5°E with heavy units taking on the cruiser escort and the smaller ships attacking the vulnerable merchantmen. On 4 July, having realized that the enemy ships had left Trondheim, and having received a report that they were close to the convoy, the Admiralty withdrew the escorts and instructed the merchant ships to scatter and proceed independently. It must have seemed to those in authority that their worst nightmares were about to become a reality. Almost immediately the German air and submarine attacks began to decimate the merchant ships, and as news of the slaughter came in Admiral Raeder decided not to endanger his powerful surface fleet and they were withdrawn when they were north of Petsamo. In the event 19 merchant ships were sunk and the equipment lost included 430 tanks, 210 aircraft and 3,350 vehicles. Worse than that 153 seamen were drowned in the action, which proved to be a major victory for the Germans, particularly as the *Tirpitz* had never come anywhere near the convoy. It was an indication of the awe which this single German battleship inspired in the Admiralty and sadly the *Victorious* and the covering force were never ordered to intervene. On Wednesday 8 July the carrier anchored in Scapa Flow where rumours abounded that she was to be dry docked.

This time the rumours were true and at 2pm on Friday 10 July 1942, flying the flag of Rear-Admiral A. J. Lister, Flag Officer Home Fleet Aircraft Carriers, she sailed for the Irish Sea and anchored close to the Mersey Bar Lightship the following evening. Next morning she steamed up the River Mersey and by midday she was secured against the North Wall of the Gladstone Dock off Bootle's Regent Road. Originally each watch was to have five days' leave, but this was reduced to three as the docking period itself was reduced to six days. However, Les Vancura recalls that within an hour of the ship securing alongside he was on his way to Lime Street Railway Station and he arrived in Swansea at 4.45am the next morning. He had to start his return journey at 9.25pm on Tuesday 14 July, arriving at Lime Street at 5.45am the next day which gave him just

A Fairey Albacore takes off from the *Victorious* whilst the ship is at anchor in Scapa Flow. *(Fleet Air Arm Museum)*

enough time for a cup of tea at the YMCA, before he got back on board at 9am to find the *Victorious* high and dry in the graving dock having her underwater hull scraped and painted.

By 3pm on Saturday 18 July the *Victorious* was clear of the Gladstone Dock and just over an hour and a half later she had crossed the bar and had set course for Scapa Flow, where she anchored at 6pm the next day. Following the disaster which befell convoy PQ17, the Arctic convoys were temporarily suspended but there was an even more urgent problem now facing the country and that was the position in the Mediterranean and North Africa. The first German troops had landed in North Africa in February 1941, under the command of General Erwin Rommel, who would earn the nickname 'Desert Fox', and his leadership qualities were acknowledged by both his comrades and his enemies alike. He had fought with bravery during the First World War on the Italian front, and between the wars he had served as a regimental commander. On the outbreak of the Second World War he was a colonel in command of Hitler's Headquarters Guard, but he was transferred in 1940 to command the 7th Panzer Division and he displayed his tactical skills during the invasion of France, leading his troops from the front with characteristic bravado. His posting to North Africa was to retrieve the situation following the collapse of the Italian Army in their campaign against Wavell's forces. Rommel soon saw that British forces in North Africa were weak, and that few reinforcements would be available. Despite some setbacks, by late May 1942 Rommel's Deutsches Afrika Korps and their Italian allies had pushed the Eighth Army back to Gazala and Tobruk and they were threatening the large British garrison there. In the Mediterranean the Royal Navy faced a difficult task since the modern, well-equipped Italian Navy enjoyed a position where they could strike at will, and despite several successes which Admiral Cunningham's Mediterranean Fleet had scored against the Italian Navy, the plight of Malta was worsening, so that by mid-1942 the British island fortress was in serious straits. Its vulnerable location astride the Axis supply lines made it the target of incessant air attacks and its own supply lines were becoming increasingly tenuous. Convoys to Malta had to be suspended in July due to the heavy losses which had been suffered. It was clear that Malta could not hold out against both the Luftwaffe and Regia Aeronautica without food or fuel, and a convoy of 14 merchantmen escorted by over 30 warships was to be dispatched in a desperate effort to convey supplies from England. This massive operation was code-named 'Operation Pedestal' and it was to include the *Victorious* as well as four other aircraft carriers.

Meanwhile, in the last week of July the *Victorious* was at Scapa Flow, putting to sea on a daily basis for exercises and to embark her squadrons. On Thursday 23 July she flew on six Fulmars of 884 Squadron and at 1pm on Friday 31 July she left Scapa Flow once again to embark the Sea Hurricanes of 885 Squadron, which was safely completed before the onset of thick fog. The ship's company were aware that the *Victorious* was bound for sunnier climes for, as Les Vancura recorded in his diary: 'We have had inoculations and have been issued with tropical kit.' Next day, as the *Victorious* steamed south through the Irish Sea, she rendezvoused with HMS *Argus*, and during the night of 2 August the convoy which, for security reasons, had been given a bogus WS number (indicating that it was a westbound Atlantic convoy), sailed from the Clyde with the cruisers *Nigeria* and *Kenya*, and a destroyer screen to join Vice-Admiral E. N. Syfret, who was flying his flag in the battleship *Nelson*, the following morning. Shortly before the departure from Scapa the Admiralty had decided that the aircraft carrier *Furious* should carry out 'Operation Bellows' to reinforce Malta with Spitfires concurrently with 'Pedestal' but, owing to technical problems with the aircrafts' propellers and the *Furious*' humped flying deck, she was unable to leave with the main body and she set sail later with the cruiser *Manchester* and steamed at high speed to join the convoy.

On Thursday 6 August the other four carriers (*Indomitable*, *Furious*, *Argus* and *Eagle*), rendezvoused west of Gibraltar and two days later aircraft from the carrier force made dummy attacks on the convoy, followed by a fly-past for identification purposes. At 11.45am on Friday 7 August, during flying operations, a Fulmar from the *Victorious* crashed into the sea with the loss of its crew, Sub-Lt C. G. Taylor RNR and Ldg Airman J. F. Elson, both from 809 Squadron. During the afternoon of Sunday 9 August, after the destroyers had refuelled at Gibraltar, the whole convoy formed up to make the passage of the Gibraltar Strait during the night. Cape Spartel was passed at midnight, and during the passage through the Strait of Gibraltar a large number of fishing boats were passed between Malabata and Tarifa, and also two neutral steamers sailing westward. However, the passage of such a huge force of ships through the narrow strait soon became known to the Axis forces and the Italian High Command immediately put in hand measures to attack the convoy as it steamed towards Malta. Admiral Syfret's dispatch sums up the convoy's passage and their first day in the Mediterranean: 'The passage of the Strait and D1 (10 August) were uneventful. Fishing boats and one merchant vessel were passed at close quarters, but aided by a moonless night and indifferent visibility it is improbable that the force was sighted from the shore. Reports received later showed, however, that the enemy was fully cognisant of our passage of the Strait.' Another interesting comment from the Admiral gives an indication of the force required to escort just 14 merchant vessels: 'When *Indomitable* joined my flag it is believed to have been the first occasion when five of HM aircraft carriers have ever operated at sea simultaneously.'

Unfortunately, day two in the Mediterranean was going to be anything but peaceful. Tuesday 11 August dawned bright and clear with every prospect of being a hot, cloudless day as the convoy steamed east between Majorca and the North African coast of Algeria. On board the *Victorious* 'Flying Stations' was sounded at 6am and the first fighter patrols were flown off, with one of the Fulmars crashing on landing, luckily with no casualties. Soon after this the first enemy reconnaissance aircraft was detected and Hurricanes from the *Indomitable* were sent to intercept it, which proved a difficult task as it was flying at 20,000 feet or more. However, one enemy aircraft was later shot down. At 10.30am one of the *Victorious*' Fulmars crashed into the sea and the aircrew were rescued by the destroyer *Wishart*. At 11.28am three distant disturbances were seen on the surface of what was a flat calm sea, which were described by officers in *Nelson* and *Charybdis* as torpedoes breaking surface. At 12.18pm HMS *Furious*, screened by the destroyers *Lightning* and *Lookout*, moved out to the port quarter of the convoy for 'Operation Bellows' and the first of 38 Spitfires took off ten minutes later to reinforce Malta's air defences. Unfortunately, one of them developed an engine fault and had to land on the *Indomitable*, but the remaining 37 arrived safely on the beleaguered island.

At 1.15pm, when the convoy was in a position Lat 38° - 05'N/Long 03° - 02'E, HMS *Eagle* was hit on the port side by a spread of four torpedoes, all within an interval of about ten seconds. At this time the *Eagle* was stationed on the quarter of the starboard wing column of the convoy and steaming at 13 knots, and within only eight minutes she had heeled over to port and disappeared beneath the waves. A lookout on one of the merchantmen recalled: 'I was astounded at the sight of the two-funnelled carrier *Eagle* apparently turning so sharply that she was heeling over.' This witness recalled a destroyer racing over to the carrier and the 'unthinkable' dawning on him as the *Eagle*, '...was turning on her side and aircraft were sliding into the sea from the flight deck.' He went on to say, 'I could not resist watching the *Eagle* but I did not see her for long because apart from the fact that she was dropping astern, she was sinking fast.' The destroyer *Lookout* which was screening the *Furious*, and the *Laforey*, together with the tug *Jaunty*, were ordered to search for survivors and between them they rescued 927 men, including Captain L. D. Mackintosh DSC RN, *Eagle's* commanding officer. Nevertheless, 160 members of her ship's company were lost in this terrible tragedy.

For about an hour and a half after the sinking of the

'Operation Pedestal', looking astern from the *Victorious* at HMS *Indomitable* and HMS *Eagle*.   *(Imperial War Museum)*

Eagle there were numerous U-boat alarms and destroyers carried out anti-submarine searches of the area, but no contacts were made. At 2.20pm enemy aircraft were detected on radar and they passed over the force at a great height a few minutes later without mounting an attack, so it was thought they were carrying out photographic reconnaissance. The *Furious* completed the flying off of her Spitfires at 3.20pm, and after transferring *Eagle's* survivors to the *Keppel*, *Venomous* and *Malcolm* she departed for Gibraltar with the three destroyers.

At 4.34pm Admiral Syfret received a signal warning him of a dusk air attack on the convoy, and radar reports clearly indicated that a raid would develop at about 9pm that evening. At just before 9pm the convoy was attacked by about 30 JU 88s and by torpedo bombers, which kept up their onslaught until about 9.30pm, but they did not cause any damage to the ships. The anti-aircraft barrage which was put up was described by Admiral Syfret as 'most spectacular' and three enemy aircraft were shot down. During the raid a number of aircraft from the *Victorious* and the *Indomitable* were airborne, but in the failing light they were unable to locate the enemy force. They then had to be landed back on after dark and in doing so some were fired on by the escorting warships. The situation was extremely dangerous and obviously the aircraft which had been airborne longest had to land first. One aircraft, a Hurricane from the *Indomitable* which was almost out of fuel, made an emergency landing on the *Victorious* despite the fact that the carrier was still turning into the wind. As the pilot landed, his plane hit another Hurricane parked abaft the island, and then skidded into the barrier where it burst into flames. Fortunately, the pilot was unharmed, the fire was soon extinguished and within minutes the flight deck was clear and ready to land on more aircraft. Finally, when all the aircraft had returned, it was found that five from the *Victorious* had landed on board the *Indomitable*. At 10.40pm, with darkness having given the convoy some respite, the *Victorious* took station astern of the battleship *Rodney* and the whole force continued to steam east at 12 knots - with the promise of a very busy day to come.

In the early hours of Wednesday 12 August, radar reports began to come in of enemy reconnaissance aircraft snooping ahead of the convoy, and at 5.20am, as soon as dawn broke, 'Action Stations' was sounded and 12 fighters from both the *Victorious* and the *Indomitable* took to the air to maintain patrols over the convoy which was by now approximately 50 miles off the coast of Algeria, north of the Algerian-Tunisian border, and some 85 miles south of Cape Spartivento in Sardinia. This was well within range of Italian airfields on the island and the enemy bombers could now be escorted by fighters. To add to his problems Admiral Syfret had also received warnings of U-boats in the area ahead of the convoy, but it was the air raids which came first that day. At just after 9am, 20 JU 88 high-level bombers approached the convoy and they were engaged by 16 fighters from both carriers, and Hurricanes from the *Indomitable* shot down two of them. These aircraft came in over the fleet, dropped their bombs, and were away in about six minutes only to be caught by the *Victorious*' fighters who shot down two more. All that morning the two carriers continued to fly patrols over the convoy, whilst the destroyers seemed to dart all over the place as they investigated asdic contacts. Admiral Syfret's dispatch states, 'Meanwhile, spasmodic firing by the screen at snoopers which came within range continued and, of course, the carriers and their "chickens" were, as always, extremely busy.'

At midday radar reports were received of an air raid which was coming in from ahead, this time perpetrated by a large force of enemy bombers. Once again, fighters from both carriers intercepted them and one bomber was soon seen to be dropping into the water in a trail of smoke and flames. It soon became apparent that this first formation of enemy aircraft was dropping parachute mines ahead of the convoy and Admiral Syfret ordered an emergency turn of 90 degrees to port to avoid them. Next came a large number of torpedo bombers which approached in formations of five or six on the port bow, port beam and starboard quarter, although none of the attacks were pressed home with determination and all the aircraft dropped their torpedoes outside the range of the convoy, with one bomber being shot down by ships' gunfire. This torpedo attack was closely followed by a large number of JU 88s which dive-bombed the convoy and a near miss alongside the steamer *Deucalion* flooded two holds and reduced her speed to 10 knots. The battleships *Rodney* and *Nelson* and the cruiser *Cairo* also experienced some very near misses. At 1.45pm the *Victorious* had a very narrow escape when two Italian Reggione fighter-bombers, which could easily be mistaken for Hurricanes, circled the carrier whilst she was landing on her own aircraft, and they were actually thought to be friendly aircraft by many of the flight deck crews. Suddenly the two aircraft dived over the *Victorious* from astern and attacked low over the flight deck releasing their bombs as they flew over the ship. Fortunately, one missed completely and exploded in the water off the port bow, and the other actually hit the carrier's armoured flight deck, failed to explode and broke into several pieces which fell harmlessly into the sea. So complete was the surprise that the enemy aircraft escaped without a shot being fired at them.

Alan Belcher of 809 Squadron remembers the incident: 'Keeping a sharp lookout for anything that was flying, we espied two aircraft in formation speeding in over the fleet towards us. From our position we assumed they were Hurricanes making a hasty land on and, as no ship was firing on them, we didn't take a lot of notice, until we saw they were making a beeline for the port side of *Victorious*. Chalky suddenly yelled, "They're not 'B' Hurricanes", whereupon a large elongated object with tail fins left the underside of the now banking aircraft and fell towards us. We all plunged face down onto the flight deck and tried to claw holes in the three-inch armour plating, and it seemed an age as we counted what we thought were our last seconds on this earth. There was a loud clang and the deck shook violently, but we were in no hurry to get up as we didn't know where the bomb had gone, but after waiting for some five or six seconds and there having been no explosion, we stood up. We then saw a nasty groove, minus paint, on the flight deck about 15 feet from us. Fortunately, the plane had been too low and the bomb had bounced and skidded along the flight deck, across the forward end of the island, before breaking into pieces and going over the starboard side.' During the attack other Italian aircraft dropped a new type of torpedo which was described as looking like a canister with a small black parachute. Meanwhile, the *Tartar* reported a submarine in sight on the starboard quarter and she carried out a depth charge attack.

During the afternoon there was only a four-hour respite between air attacks, but there were innumerable reports of submarine sightings and Admiral Syfret ordered the destroyers to drop depth charges on each side of the convoy screen every ten minutes until darkness fell. At 4.49pm the destroyer *Ithuriel* sighted a periscope and part of a conning tower on her starboard bow and launched an immediate attack. The pattern of depth charges brought the U-boat to the surface and *Ithuriel* opened fire, turned and finally rammed the submarine which turned out to be the Italian Navy's *Cobalto*. Although she was clearly sinking, a boarding party was sent over from the *Ithuriel* and before she sank, three officers, including the CO, and 38 ratings were taken prisoner. Meanwhile, reports of small formations of enemy aircraft were coming in and once it became evident that a heavy air attack was imminent every available fighter was scrambled from the two carriers yet again. At 5.36pm the fighters made contact with enemy formations and it soon became clear that not only was there a considerable bomber force on its way, but that they were also escorted by fighters. Despite a furious defence by the aircraft from *Victorious* and *Indomitable* formations of bombers broke through and at 5.50pm the *Ithuriel* was attacked, before the convoy altered course to pass through the Skerki Channel, and at 6.30pm between 100 and 120 enemy bombers and fighters were over the convoy. Against this massive force there were just 22 Fleet Air Arm fighters in the air, but despite the overwhelming odds they continually harassed and broke up the enemy formations. The first major air attack started at 6.35pm and this comprised of at least 13 torpedo bombers and a large number of high-level bombers, dive-bombers and minelaying aircraft. The whole convoy made an emergency turn to avoid mines and torpedoes which had been dropped ouside the screen, and very soon after this 40 torpedo bombers were reported ahead, followed by a JU 88 attack on the *Indomitable* which became totally obscured by bomb splashes and thick black smoke.

One officer on board the *Indomitable* recalled: 'I caught sight of a closely bunched formation of 12 Stukas at about 10,000 feet almost directly above us with not a shot being fired at them. The first Stuka peeled off and I had my first head-on view of its sinister cranked wings heading straight towards us with others following behind. I felt disembodied and without fear, unable to move - not that there was anywhere to move to. Looking like beer barrels, the 1,000lb bombs seemed to float down towards us as though in some dimly remembered dream. The ship shuddered and the dream expanded into a huge sheet of flame which rose up ahead of the island and engulfed it. There was an enormous explosion just ahead of me and then several behind me which seemed to lift the ship several feet. A wall of water rose alongside to some hundred feet then cascaded down on top of me, washing me into the catwalk. For a moment there was a strange silence as flames and smoke billowed near the forward lift and behind the after lift.'

The *Indomitable* had been hit by three large bombs and there had also been several near misses. Two large fires were raging and with the flight deck out of action and listing to port, she reduced speed and turned to the west away from the wind. The two armour-piercing bombs had caused severe damage to the flight deck forward and aft and one of the near misses had blown a 30ft hole in the port side of the hull aft, which virtually destroyed the wardroom and killed six aircrew who were there at the time. Admiral Syfret ordered the *Charybdis* and destroyers to form a screen round the damaged carrier and one of them, HMS *Lookout*, began hosing water into the gaping hole in the hull.

Meanwhile, the *Indomitable's* fighters, which were

The *Victorious*, *Indomitable* and *Eagle* (foreground) during 'Operation Pedestal'. *(Imperial War Museum)*

heavily engaged with the enemy attack, were unable to land on their own carrier and at 7pm, when the attack was over, they had to land on board the *Victorious* and by 7.30pm, apart from one Martlet which crashed into the sea, all the aircraft had landed safely on. Fortunately, the *Indomitable* was soon steaming at 17 knots and by 8.30pm her engineers could get 28 knots out of the badly shaken machinery. That evening Vice-Admiral Syfret ordered Force Z, the battleships *Rodney* and *Nelson*, the aircraft carriers *Victorious* and *Indomitable*, the cruisers *Phoebe*, *Sirius* and *Charybdis*, together with the 19th Destroyer Flotilla, to turn round and set course for Gibraltar. There was not enough fuel at Malta to replenish the capital ships, and of course there was a danger that the *Victorious* would be disabled as the *Indomitable* had been. At that stage there was still reason to suppose that the bulk of the convoy would reach Malta's Grand Harbour safely. Unfortunately, less than an hour after the detachment of Force Z, the cruiser *Nigeria* was damaged by a torpedo strike, followed closely by the cruiser *Cairo* and the two merchant ships *Ohio* and *Brisbane Star*. The *Nigeria* and the two merchant ships were able to get under way again, but the *Cairo's* stern had been blown off and as soon as all the survivors had been taken off she was sunk by other ships of the escort. Later that evening there were further air attacks on the convoy which damaged several merchant ships and two, *Empire Hope* and *Glenorchy*, were sunk. At just after 11pm the cruiser *Kenya* was damaged by a torpedo hit and half an hour later SS *Deucalion*, which had been proceeding independently following damage received in an earlier attack, was torpedoed and sunk.

During the early hours of Friday 14 August, as the convoy steamed slowly past Cape Bon in Tunisia, E-boats were detected by radar on the port beam and for three hours there were running fights as they continued to attack the convoy. At least one, and possibly two, of the marauders were destroyed, but they managed to cause heavy losses to the convoy with the cruiser *Manchester* being torpedoed and three merchant ships, the *Santa Eliza*, *Almeria Lykes* and *Wairangi* were also lost. Soon after dawn SS *Waimarama* was sunk and the escorting destroyers picked up 150 survivors from HMS *Manchester*, but several hundred of the cruiser's ship's company were taken prisoner by the Vichy French authorities in Algeria. Later that day and early on 15 August, just six of the merchant ships made it to Malta, including the gallant American oil tanker *Ohio*, which entered Grand Harbour at 8am escorted by the destroyers *Penn*, *Bramham* and *Ledbury*.

After leaving the convoy at 7pm on Wednesday 12 August, the *Victorious* and other heavy units of Force Z steamed hard for Gibraltar and at 2pm on Saturday 15 August two Albacores were launched, one of which was transporting Rear-Admiral Lyster to Gibraltar. Five and a half hours later, at 7.30pm, the *Victorious* secured alongside No 52 berth on the detached mole of Gibraltar Harbour where she embarked 501 officers and ratings from the sunken ships *Eagle*, *Manchester*, *Foresight* and the badly damaged *Indomitable*, for passage back to the UK.*

Early next morning, at 3.45am, the *Victorious* slipped her moorings to return to Scapa Flow, but whilst she was being manoeuvred from the berth by tugs she collided with the *Furious* which was tied up at the detached mole, causing some minor damage to her bows. However, this did not hold up her passage and at 6,20pm on Friday 21 August she anchored safely in Scapa Flow. Les Vancura recalled the carrier's reception on her return: 'We arrived at Scapa and manned ship, being cheered as we passed the *Kent*, *Renown*, *Anson*, *King George V* and the *Suffolk*. The *Renown's* Royal Marines Band was there playing "Hail The Conquering Hero Comes" and we all felt a bit sheepish.'

'Operation Pedestal' had succeeded in taking the pressure off the besieged island of Malta, but it was fortunate that the Allies were about to go on the offensive in the Mediterranean for the losses suffered escorting this single convoy were extremely heavy, and could not be repeated.

In his dispatch Vice-Admiral Syfret wrote: 'The work of the aircraft carriers (HMS *Indomitable*, Captain T. H. Troubridge, and HMS *Victorious*, Captain H. C. Bovell) under the command of Rear-Admiral Lyster, was excellently performed, while that of their fighters was magnificent. Flying at great heights, constantly chasing the faster JU 88s, warning the fleet of approaching formations, breaking up the latter, and in the later stages doing their work in the face of superior enemy fighter planes, they were grand. The fact that 39 certainties were shot down by them and the probability that at least the same number were incapacitated is a remarkable measure of the success of the carriers, their teamwork and their fighters and of the able and inspiring leadership of Rear-Admiral A. L. St G. Lyster CB CVO DSO.' He also praised the flight deck personnel thus: 'Nor must the work of those who kept the fighters in the sky be forgotten. The flight deck parties, composed of Able Seamen, had been on duty from dawn till dusk ranging the aircraft, placing the chocks in position and removing them on the signal, releasing the landing aircraft's hooks from the arrester wires and folding the wings. On day three the party in the *Victorious* made 86 journeys up and down the flight deck, a distance of something like 20 miles. In the hangars, "The maintenance ratings worked like slaves in the cramped space. The decks were covered with grease and running with oil and petrol. The heat was intense and they had to work all the time under electric light, taking their food in the hangars as and when they could". This is the tribute of a pilot in the *Victorious* to those in whose devoted service the fighters did not trust in vain.'

---

* The survivors were made up as follows: 25 officers and 157 men ex-HMS *Eagle*; three officers and 132 men ex-HMS *Manchester*. Seven officers and 132 men ex-HMS *Foresight*: ten officers ex-HMS *Indomitable*.

Chapter Five

'Operation Torch' And The Pacific

Just 24 hours after her return from 'Operation Pedestal' the *Victorious* was at sea again, this time for a fast overnight passage to the Naval Dockyard at Rosyth, where she arrived at just after 9am on Sunday 23 August. Two days later she moved into No 2 dry dock and, as all the mess decks were to be repainted, those members of the ship's company who had not gone on leave moved into shore accommodation. During the refit two self-contained diesel-generators were fitted, additional escape hatches and fire-fighting equipment were also installed and Les Vancura spent a very strenuous day embarking spare parts for Seafires, which was a good indication of the type of aircraft which were to be embarked. At 6.30pm on Thursday 17 September *Victorious* left the dry dock and after passing through the dockyard locks she anchored in mid-river an hour later. Over the next two days stores and ammunition were embarked, and at just before midday on Sunday 20 September she sailed for Scapa Flow where she arrived the following morning. During the remainder of the month she put to sea a number of times to carry out various flying and deck landing trials, initially with the US-built high-performance Grumman Martlet (more appropriately named Wildcat in the USA), and the distinctly odd looking Fairey Barracuda. The Martlet was, without doubt, the best naval fighter available at that time and it was supplied under the Lend-Lease agreements. With a top speed of 280 knots it surpassed the Fulmar and, up to 15,000 feet, was faster than the Hurricane too. Although it was slightly slower than the Seafire its endurance at full speed was two and three-quarter hours, compared to the Seafire's three-quarters of an hour. It was armed with four ·50mm calibre Colt machine-guns which, although they had a slower rate of fire, were considered superior to the Fulmar's eight ·303 machine-guns. The original prototype of the Barracuda had carried out trials on the *Victorious* in May 1941, following which it had been fitted with a new tail unit and was known thereafter as the Barracuda II. However, with its large flaps which were designed to give additional wing area and with what looked like a large television aerial above its port wing, which was, in fact, long-wave ASV radar, it was still a real oddity. One Fleet Air Arm pilot who flew the machine remarked, 'It was underpowered and obsolete before it went into service.'

During the first two weeks of October the *Victorious*

HMS *Victorious* at anchor in late 1942. (C. Minett)

continued her flying trials with Seafires, Martlets and Barracudas and on 11 October, whilst at anchor, she received a visit from the Prime Minister, Mr Winston Churchill, and the Lord Privy Seal, Sir Stafford Cripps, who were visiting the fleet at Scapa Flow. Over the following three days there were three Seafire crashes, two on deck whilst landing and one whilst taking off, when the aircraft fell into the sea off the port bow. At 4.30pm on the afternoon of Wednesday 14 October 1942, Captain L. D. Mackintosh DSO DSC RN, joined the ship. Although he would not take over command from Captain Bovell for over five weeks, he would gain valuable experience during the

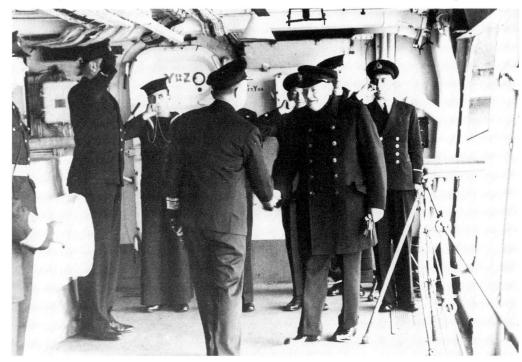

On 11 October 1942 the Prime Minister, Winston Churchill...

...and the Lord Privy Seal, Sir Stafford Cripps, visited the *Victorious*.
(C. Minett)

next major Mediterranean operation, the invasion of Vichy French North Africa code-named 'Operation Torch'.

Early 1942 had seen some serious setbacks for the Allies, particularly in the Far East where the British garrison of 90,000 troops surrendered at Singapore in what has been described as the 'greatest disaster to befall British arms', and the American forces in the Philippines surrendered to the Japanese. As the Japanese Army advanced into Burma and their naval aircraft raided the British Naval Base at Trincomalee, the Indian Ocean became a very dangerous area for Allied shipping. At the highest level it was decided that the maximum Anglo-American effort should be put into the defeat of Germany, and with the political pressure for a second front it was clear that the Allies would have to take the initiative somewhere. All through 1942 the action in North Africa, with British and Commonwealth forces on the one hand and the Axis forces on the other, seemed to sway first one way and then the other. In the autumn of 1942 it was decided that there would be a decisive Allied effort to finish the North African campaign, which would be a prelude to an invasion of Italy. A landing in the Vichy-held countries of Algeria and Morocco would not be without its dangers, for it could quite easily bog down the Allies in a campaign similar to Gallipoli during the First World War, which has been described as 'a conflict in the wrong place, at the wrong time and against the wrong enemy'. The Allies' objectives were to deny the control of West Africa to the Axis Powers and to provide a base for the eventual operations against southern Europe. The invasion itself was code-named 'Operation Torch' and there were to be three separate assault points for the invading forces. The western task force was to land in Morocco, the central task force at Oran, while Algiers was the landing place for the eastern task force.

The *Victorious* left Scapa Flow at 9am on the morning of Friday 16 October and steamed south to Greenock, where she arrived the next morning to find the great Cunarder *Queen Mary* at anchor, together with a number of other large passenger liners which were all serving as troop transports. For the next 13 days the *Victorious* sailed each morning from her Greenock anchorage to carry out various flying exercises and to provide air cover for Army landing exercises on the shores of the island of Bute. She undertook deck landing trials with the two newly operational Lend-Lease carriers, *Avenger* and *Biter*, both of which were equipped with Swordfish and Sea Hurricanes. After spending two days in the Irish Sea, during the early hours of Saturday 24 October the *Victorious* anchored in Belfast Lough and after daylight she returned to Greenock where Les Vancura noted in his diary, 'We are chock-a-block with planes and spares, with main planes being stowed on the hangar sides. We are carrying 61 aircraft at the moment, 38 fighters and 23 torpedo bombers'. It was apparent to everyone on board the *Victorious* that something big was in the offing and that they were to be part of it. Les Vancura wrote in his diary: 'Troop carriers are gathered here, all full, so it looks like a landing of some sort. Stores have been coming in all day and we are taking on stores for *Biter*, *Dasher* and *Argus*.'

On Sunday 25 October, 39 large passenger liners, all of which had been converted for service as troop transports, left the Clyde to steam south-west well into the Atlantic, before curving south-east for the Strait of Gibraltar. The seven aircraft carriers which would provide the main air cover for the central and eastern landings were *Victorious*, *Formidable*, *Argus*, *Furious*, *Avenger*, *Biter* and *Dasher*, and the *Victorious*, flying the flag of Rear-Admiral A. L. St G. Lyster, was to form part of Force H with the *Formidable*, *Argus* and *Avenger* in the operation to cover the eastern landings at Algiers and Bougie. Escorted by the *Nelson*, *Renown* and *Duke of York* together with cruisers and destroyers, Force H left the Clyde during the evening of Friday 30 October and set course for Gibraltar, sailing via the west coast of Ireland. Embarked in the *Victorious* were 809 (Fulmar), 882 (Martlet), 884 (Seafire) and 817 and 832 Albacore Squadrons. On the evening of Thursday 5 November, with Force H off southern Portugal, the *Duke of York*, *Renown* and five destroyers detached to Gibraltar for refuelling while the remainder of the force stood off to westward, out of sight of the coast. During the passage the hangars were a hive of activity as US markings were painted on all the aircraft, over the traditional British roundels. Elsewhere, in the passageways preparations for action damage were made, with lines being rigged to allow for easier access should the ship take on a heavy list and food dispenser points being set up in case the carrier suffered hits to her main galley areas. It was hoped that the Vichy French forces would offer little or no resistance to US forces and the American Government had undertaken great diplomatic efforts to try to ensure this. Unfortunately, Anglo-French relations were still embittered by the events of 1940 and particularly by the Royal Navy's attack on the French Fleet at Oran in the summer of that year.

During the early hours of Friday 6 November the massive convoy of troop transports and all the naval warships passed safely through the Strait of Gibraltar but, like the 'Pedestal' convoy of just three months earlier, it did not go unnoticed by the Axis agents in southern Spain. At 5.15pm that afternoon, when Force H was in a position Lat 36° - 51'N/Long 00° - 38'E, a fighter from HMS *Formidable* shot down an Italian reconnaissance aircraft which had been shadowing the convoy. During the afternoon of Saturday 7 November, as the convoy approached Algiers, fighters from the *Victorious* and *Formidable* engaged and broke up bombers which had been heading for the troop convoy, although the destroyer *Martin* suffered a near miss during the attack. Early on Sunday 8 November 1942 the first assault forces landed close to

Here HMS *Victorious* is shown in the Pacific Ocean during 1943, with Martlets on the flight deck. *(C. Minett)*

Algiers, and fortunately there was little resistance as the troops established their bridgeheads and started to move inland. As soon as dawn broke, the *Victorious* and *Formidable*, which were standing off the coast about 40 miles from the city of Algiers, flew their first air patrols. At just before 6am four Martlets of 882 Squadron were launched from the *Victorious* to cover the French airfield at Blida, which was some 20 miles south-west of the city of Algiers, and soon after this a stronger force from both carriers was patrolling over the airfields at Blida and Maison Blanche, ten miles south-east of Algiers, to prevent any French aircraft from taking off. They actually destroyed two aircraft which were preparing to get airborne and a later patrol succeeded in capturing the Blida airfield base. Lieutenant B. Nation, who was leading the patrol, noticed that white flags were being waved and he was granted permission to land, whereupon he accepted the surrender of the camp's commandant. That same day the Albacores of 832 Squadron were in action when they attacked French forts, the first on a breakwater of Algiers Harbour and a second which was situated on a defensive escarpment outside the city. Both attacks were carried out successfully, as were similar raids by the Albacores from the *Formidable*.

Next day at just after 1pm, two enemy aircraft were intercepted by fighters from the *Victorious*, with one of them being shot down and although the second managed to escape he jettisoned his bombs. That afternoon, at 5.15pm, a formation of enemy JU 88s were intercepted by Martlets of 882 Squadron and once again they were prevented from attacking Force H. In the early hours of Tuesday 10 November, just when everything seemed to be going smoothly for the Allies, the destroyer HMS *Martin* was torpedoed by a U-boat and she sank with heavy loss of life. Three days later the U-boats scored another success when, at 5.30am on 13 November, the Dutch destroyer *Isaac Sweers* was sunk. It was reported that there were at least 20 enemy submarines in the invasion area off North Africa during 'Operation Torch' and they took a heavy toll of both troop transports and warships, and on Sunday 15 November, *U115* commanded by KLt Adolf Piening, torpedoed the brand new escort carrier *Avenger* just west of the Strait of Gibraltar. The torpedo hit the ship and exploded abreast her bomb room and the enormous explosion broke her back, causing her to sink within five minutes with the loss of over 500 lives. Fortunately, the *Victorious* berthed safely that evening alongside the detached

47

mole at Gibraltar, where she remained for three days. Although 'Operation Torch' was executed successfully, the failure of the Allies to capture Tunisia quickly proved to be a weak point and four large troop transports were lost to either U-boats or air attacks.

At 4pm on Wednesday 18 November 1942, with the RAF established in Algeria, the *Victorious* left Gibraltar with the battleship *Duke of York* and a destroyer screen to return to Greenock. Three days out of Gibraltar, as the force was crossing the Bay of Biscay, an Albacore of 817 Squadron scored a notable success. The aircraft, piloted by Sub-Lt Thomas H. Hands RNVR, and crewed by Sub-Lt William O. Findlay RNVR and Ldg Airman Edwin F. H. Hartnell, had taken off on an anti-submarine patrol on the morning of Saturday 21 November when, at 10.40am, they spotted a U-boat (*U 517* commanded by KLt P. Hartwig) on the surface in a position Lat 46° - 15'N/Long 17° - 10'W, at a distance of about two miles. The aircraft immediately closed to attack from astern and at the same time the U-boat began to submerge and disappeared beneath the waves when the aircraft was still some 500 yards astern of it. However, as the Albacore passed over the spot where she had last been seen, four depth charges were dropped. All four were seen to explode and after two minutes the U-boat's bow broke surface at a very steep angle, and it was evident to the Albacore's crew that she was badly damaged. The destroyer *Opportune* was sent to close the position and to search the area where she soon found that the enemy submarine was sinking rapidly and all that was left for the destroyer to do was to pick up 51 survivors and rejoin the force, which she did at 1pm. A member of HMS *Opportune's* ship's company recalls the incident: 'Eventually we located the U-boat. We were at "Action Stations" and our skipper's intention was to take the sub in tow back to the UK. As we approached, machine-guns were fired over the conning tower and the crew began leaving the U-boat in record time. Our whaler was soon launched and they were rowing towards the enemy sub when she began sinking rapidly, obviously scuttled. The U-boat's crew were all in the water singing their national anthem and I remember our skipper shouting over a loudhailer, "Leave them in the water until they stop singing!" We eventually picked up 51 survivors including KLt Paul Hartwig, who retired from the German Navy in 1975 as a Vice-Admiral and C-in-C of the Fleet.'

Two days later, at noon on Monday 23 November, the *Victorious* tied up to the flagship buoy at Greenock and Captain Mackintosh took command. The rest of the ship's company could now collect their mail and for those who had not been able to take leave in September there was a welcome break ahead. Les Vancura was obviously wondering where the carrier's next destination was to be for he noted in his diary, 'Putting all British aircraft spares ashore and taking on Martlet spares.' All would soon become clear, but first of all, on the afternoon of Saturday 28 November, Captain Bovell took his leave of the ship he had seen completed on the River Tyne and which he commanded for the first 19 months of her very eventful career. He departed in some style with the Royal Marines Band playing 'Auld Lang Syne' while Heads of Department rowed him to the shore.

Captain Mackintosh had joined the Navy as a midshipman in 1914 and had been awarded the DSC during the First World War. In 1922 he had specialized as a Naval Air Observer and three years later he had qualified as a pilot, having taught himself to fly. He was promoted Captain in 1938 and he had commanded the cruiser *Charybdis* in the early months of the Second World War. In early 1942 he was appointed to command the aircraft carrier HMS *Eagle*, and he was in command of her when she was torpedoed and sunk during 'Operation Pedestal'. He was one of 535 survivors who had been rescued from the water by HMS *Lookout* and then he joined the *Victorious* on 14 October 1942.

During the first two weeks of December the *Victorious* remained at Greenock, putting to sea on a daily basis for gunnery exercises or to land on aircraft. There were indications that her next deployment was not to be with the Home Fleet for as Les Vancura recorded in his diary for 15 December, 'All RAF personnel left ship and also Admiral's staff, this means we are leaving the Home Fleet.' Two days later he was able to record, 'A batch of ratings embarked for transit and they seem to be bound for the USA. We now have 40 Martlets on board and rumour has it we will be having a refit while we are away.' Les was quite right in his assumptions for the *Victorious* was bound for the USA to join the US Pacific Fleet which was short of aircraft carriers. In the months following the Japanese attack on Pearl Harbor the US Navy had lost four aircraft carriers in the Pacific war; the *Lexington*, during the Battle of the Coral Sea, the *Yorktown* at the Battle of Midway, the *Wasp* in the fierce battles which took place for the island of Guadalcanal and the *Hornet* during the Battle of Santa Cruz, which was the third major Japanese attempt to retake Guadalcanal. During the latter battle the USS *Enterprise* had also been seriously damaged, which left her with much reduced fighting efficiency. As a result of these heavy losses Admiral Halsey, who was in command of US Naval Forces in the South Pacific, requested that a British aircraft carrier be sent to reinforce the US Navy in an effort to alleviate his desperate shortage of aircraft carriers in the area.

At just after 2pm on Sunday 20 December 1942 the *Victorious* slipped her mooring at Greenock's flagship buoy and, together with the destroyers *Racehorse* and *Redoubt*, she set course for Bermuda. Two days later, when the carrier was some 300 miles west of Ireland, she ran into a severe Atlantic storm with winds of over 56 knots, and with many cooks suffering from seasickness Les Vancura recalls how he helped to carry the Christmas puddings from the stores up to the galley. On Christmas Day the wind speeds dropped to 41 knots and a full Christmas dinner, complete with roast turkey, ham, stuffing, roast potatoes, peas and parsnips, was served. For tea there was pineapple and fruit cake, and for supper cold turkey, ham and mince pies, followed by some with a sing-song and games in their mess. Fortunately, by Boxing Day the storm had abated somewhat and anti-submarine patrols were able to fly once more. At 8.30am on Sunday 27 December, the destroyer *Redoubt* came alongside to refuel, a manoeuvre which was made particularly difficult by the heavy swell which was still running. Eventually the destroyer was able to take station alongside the carrier and the fuelling hoses were connected, but not for long. At just after 9am the destroyer was carried by the sea right into the starboard side of the *Victorious*, which resulted in some damage to the carrier's lattice radio masts. Fortunately, the *Redoubt* was able to steer clear, but not before the fuelling hoses had parted and had been carried away. Needless to say the attempt at refuelling was abandoned. As the force neared Bermuda and the weather got much warmer, the storms died away as well and the damage to the carrier could be assessed. Not only were the starboard after radio masts stuck in the up position (which was very disconcerting for pilots landing on), but there was also some damage to the bow plates. However, up on the flight deck the Martlets which had been fastened down for the voyage were found to have weathered the storm without suffering any damage at all. Finally, at just before 9am on Wednesday 30 December, the *Victorious* anchored off Bermuda where the sea was like a mirror, and with water temperatures of just over 60°F, 'Hands to Bathe' was piped. Bermuda was at that time a US base under the Lend-Lease agreements and so the *Victorious* was something of a curiosity for the American service personnel based there. The visit was short-lived though, and at 4.35pm the same day she weighed anchor, bound for the US Naval Base at Norfolk, Virginia.

En route to Virginia Les Vancura recalls that the Supply Officer gave the ship's company a talk on the 'mysteries' of the US dollar since, after their arrival in Norfolk, they were to be paid in American currency. At 10am on New Year's Day 1943, the American coast was sighted for the first time and soon afterwards the *Victorious* entered the Hampton Roads, passing the site of that historical naval battle of 1862 between the first ironclads *Merrimac* and *Monitor*, before entering the Elizabeth River for Norfolk where, at 5.50pm she secured alongside a basin in the US Naval Dockyard. Les Vancura has recorded the event in his diary thus: 'Proceeded upriver, passing Norfolk and Portsmouth, we lined ship with the Royal Marines Band playing American tunes. We were cheered from the ferries as they passed us and after three miles we moored at the Naval Yard, not far from the carrier USS *Essex*.' It was a good start to what was going to be a very enjoyable stay for the ship's company, and a flavour of this and of the very generous welcome offered to all those on board by the people of Norfolk can be gathered from more of Les Vancura's memories: 'Marvellous conditions ashore. It seems so strange to have no blackout and we have been given ten dollars each to tide us over until we can change our English money. There is a grand canteen in the dockyard with a different film show each night and admission only a dime. We can get fresh milk in waxed cartons, honey-dipped doughnuts and milk chocolate from a kiosk on the jetty. Marvellous hospitality from the local people. Len, Ron and I went ashore, had "big eats" in the Southern Restaurant, strolled around the shops and went to see "Life Begins At 8.30" starring Ida Lupino, before having supper in a drugstore and back on board by 11.30pm.' Two days later he went ashore again: 'I took a bus to a small country place outside Norfolk, called Suffolk. I strolled around and bought two pairs of sunglasses where the manager of a "help yourself" shop showed me round the place. Saw Judy Garland in "Me and My Girl". Back on board by 11.30pm.' For the *Victorious*' ship's company the USA was a whole world away from the austerity of wartime Britain and many took advantage of coach tours to Washington and New York, while the pilots and aircrew of the squadrons had to familiarize themselves with the US Navy's aviation practices and terminology. There were some difficulties with the American system of landing signals which, in some cases, had very different meanings. The *Victorious* herself underwent a ten-day period in dry dock and on Thursday 21 January 1943 she was shifted to a berth at Pier 3 in the US Navy Yard for ammunition and stores to be embarked. Some alterations had been made to the ship's appearance during this time, with gun sponsons being fitted, together with Oerliken anti-aircraft guns, round the stern. The hull was spray painted with camouflage and the hangars were crammed with spares for the aircraft. Finally, at 9am on Saturday 30 January 1943 the *Victorious* slipped her moorings and put to sea for her post-refit trials. It had been the most extensive refit which the ship had undergone since she was first commissioned almost two years previously.

The *Victorious* was now part of the US Navy's powerful Pacific Fleet and to other units of the US Navy this strange 'Limey flat top' became known as the USS *Robin*. One of the most popular adaptations was the use of the US Navy's working dress which consisted of light blue cotton shirts and dark blue jeans for the ratings and khaki for the officers. So popular were these garments with all the ratings on board the *Victorious*, and so practical were they when compared to the blue serge and overalls which the men were used to wearing, that a recommendation was made to the Admiralty that it be adopted by the Royal Navy as a new Action Working Dress. With only some minor changes the new uniform became standard issue in the Royal Navy and it remains today of course as No 8 AWD and is now worn by officers and ratings alike. After a day at sea the *Victorious* returned to the Norfolk US Navy Yard on 31 January where she secured at No 7 Pier for a further two days before leaving on Tuesday 2 February to undertake exercises and to embark her squadrons. For four days the ship's company carried out various exercises during which they learned that 'Abandon Ship Stations' was now known as 'Boat Stations'. Aircraft were embarked and on Saturday 6 February, after weighing anchor at 8am, the *Victorious* left Hampton Roads and set course for Colon and the Panama Canal. After two days, as the weather got noticeably warmer, the lights of the Bahamas became visible and the following day she passed the island of Cuba. Next morning, at 7.45am, whilst flying anti-submarine patrols, a Martlet crashed into the sea and sadly the pilot, Sub-Lt T. Hutchinson RNVR, was killed. That same day another Martlet crashed into the sea, but fortunately this time the pilot was rescued. By now the ship's company had changed into tropical rig and at 1.45pm on Thursday 11 February the carrier secured alongside Pier 6 at Colon. The ship's company could now enjoy open-air cinema shows in the open lift wells and unlimited supplies of fresh fruit were brought on board. Les Vancura has noted in his diary that, although there was no leave granted, apples and oranges could be bought on the dockside at 5 cents for six, and the iced-water fountain was well frequented. Meanwhile, on a more serious note, local dockyard workmen came aboard to begin the task of cutting away gun sponsons on both sides of the ship so that she would be able to pass through the locks of the Panama Canal. Finally, in the early afternoon of Saturday 13 February, the Panama Canal pilot was embarked and the *Victorious* slipped her moorings for her transit of the waterway. At 4.30pm she entered Gatun Locks and just over an hour later she anchored for the night in Gatun Lake. Next morning, at 6am, she weighed anchor to begin her passage through the Gaillard Cut to the Pedro Miguel Locks at the southern end, where she was lowered 31 feet in one step to the artificial Miraflores Lake which separates the two sets of Pacific locks. Finally, at 1.45pm, she cleared the Pacific locks and an hour later she secured alongside berth 15BC at Panama close to the battleship USS *Massachusetts* whose band played 'God Save The King' as the *Victorious* moved to her berth. Next day shore leave was granted and Les Vancura took a bus into Panama City where he found everything in the shops to be very expensive. However, in the US Navy canteen at Balboa bananas were 50 cents a stalk and, as always, ice cream of just about every flavour was cheap and plentiful. After a four-day stopover, during which time the gun sponsons were welded back into place, the *Victorious* left Panama bound for Pearl Harbor. During the long Pacific voyage she was escorted by three brand new Fletcher-class destroyers, *Bache*, *Converse* and *Pringle*, all of which were determined to look after their new and unusual 'flat top'.

Early in the voyage the *Victorious* got some idea of the massive resources and efficiency of the US Navy when a rating on board fell seriously ill with diptheria and all the serum in both the carrier and her escorts had been used up. Breaking radio silence a request for further supplies was made, and within 24 hours the US Command had a Liberator airborne and over the carrier with more of the precious serum which was dropped in a watertight container and thus saved the man's life. Throughout the passage the pilots and aircrew continually practised the US Navy style landings, which taxed them and the Batting Officer, Lt J. G. (Tommy) Thomas, to the limit. Inevitably, with all the completely new routines which had to be learned it was not long before there were casualties and on Monday 22 February, at 10.15am, a Martlet crashed whilst landing. Just over 30 minutes later an Albacore crash-landed and carried a member of the maintenance party, EA Bennett, overboard. Sadly, his body was never found. Two days later, at 4pm on Thursday 25 February, an Avenger of 832 Squadron crewed by Lt Eyre RNVR, Sub-Lt Browne RNVR and Ldg Airman Lovell, caught fire and burnt out. The incident is described here by Lt J. W. Herbert: 'There was one bad crash when a TBF caught a wire but trickled over the side into the catwalk. The propeller cut a high-pressure petrol pipe and

On 29 April 1943, whilst the *Victorious* was at Pearl Harbor, she was visited by Admiral Chester W. Nimitz, the C-in-C Pacific Ocean Areas. In this view Admiral Nimitz is welcomed aboard by Captain Mackintosh.   (*C. Minett*)

there was an enormous fire. Nothing was left of the aircraft except the engine. The three members of the crew were wearing shorts and they were all very badly burnt. They died some days later, but I will never forget the sickly smell of burnt flesh which permeated the whole ship.' The fire actually burnt out a motor cutter and a motor boat, together with timber which was stored on the boat deck sponson for damage control purposes. Initially the fire parties used water on the petrol fire, but fortunately they were soon ordered to use foam, and it was very fortunate that the *Victorious* had an armoured flight deck and not a wooden one, which would have had appalling consequences. Sadly Lt Eyre and Sub-Lt Browne died from their injuries the next day and their funeral service was held that evening. On the morning of Saturday 27 February another Martlet crashed whilst landing, causing a small fire, but fortunately it was soon extinguished and the pilot was not injured. However, on the evening of Sunday 28 February, Ldg Airman Lovell died from the injuries he had received three days earlier and his funeral service was held on the quarterdeck the following evening. During the last 48 hours of the voyage night encounter exercises were carried out, and Les Vancura recalls that a US Navy officer briefed the ship's company on the Naval Base at Pearl Harbor and told them that the *Victorious* would be berthing at Ford Island in the middle of the harbour, which meant a ferry ride to Pearl City for liberty men. He also gave them the bad news that blackout regulations were in force and that all leave expired each evening at 6pm.

    Landfall was made off Oahu at 10am on Thursday 4 March and after steaming past Honolulu the pilot was embarked at midday, and just over an hour later the *Victorious* was secured alongside 16 berth, Ford Island, Pearl Harbor, close to the US Navy airfield. As she docked American sailors crowded the shore to get a look at the 'Limey' carrier and a US Navy band played the traditional Hawaiian welcoming music before breaking into more familiar British melodies. Once ashore the visitors could sample the delights of the US Navy canteen, complete with a beer garden, swimming pool and a well-stocked shop. A trip into Honolulu was seven miles, with Waikiki Beach being a further two miles, and Les Vancura managed to borrow a surfboard to try out the legendary Pacific rollers which crashed onto the beaches of Hawaii.

The US Navy's superb cinema showed two movies every day with a different show each night, including such favourites as 'The Road to Morocco' starring Bob Hope and Bing Crosby. Apparently the Royal Navy's tropical rig of white shorts caused some amusement among the US sailors who wore white bell-bottomed trousers with their cooler and more practical short-sleeved, open-necked shirts. John Herbert, who was part of the aircrew, recalls those long, hot days at Pearl Harbor: 'We had a good time in Hawaii - the fighter squadrons were ashore at USNAS Barbers Point. We got to see quite a lot of the island and were treated to superb hospitality by local families. We managed to spend time on the beautiful beach and I also learned the joy of gin and fresh pineapple juice. I will always remember flying through the smoke from burning sugar cane - it smelled just like delicious home-made toffee. There were a couple of rather close chimney stacks at Ewa and one day we flew between them in line astern. The Americans were nice enough not to notice the incident.'

Also very much in evidence at that time were the massive salvage operations being carried out in Pearl Harbor itself to raise the sunken warships, including the battleship *Oklahoma*. On the morning of Friday 5 March the *Victorious* was visited by Admiral Chester W. Nimitz, the C-in-C of the Pacific Ocean Areas Command, and dockyard workmen swarmed aboard to repair the fire damage of 25 February and to modify the arrester gear in order that it could cope with the heavy Avengers of 832 Squadron. After five days alongside the *Victorious* put to sea for a day in order to embark her Avengers and Martlets, but by 5.15pm she was back at 16 berth. On board the carrier all the inflammable deck coverings were now removed, together with any rugs and carpets from the wardroom and cabins, and the ship was being painted once again, this time in plain uniform grey. On a lighter note, three ice cream dispensers and a Coca-Cola vending machine were installed at this time.

On Monday 12 April the *Victorious* went to sea for three days of deck landing trials and gunnery shoots, and after returning to Pearl Harbor for a long weekend she went to sea again on Monday 19 April for a further five days of exercises. During the afternoon of Thursday 29 April Ceremonial Divisions and march past were held on the flight deck, with the salute being taken by Admiral Nimitz. On Monday 3 May, the *Victorious* put to sea again for further flying trials which lasted until 7pm the next day when she berthed once again alongside the Ford Island Navy Yard. The carrier's final full day at Pearl Harbor was a very busy one, particularly for the squadron personnel who spent most of the day loading aircraft from the dockside. There was also an air raid alert which turned out to be a false alarm but, with the events of 7 December 1941 in mind, it was no surprise that everyone closed up at their Action Stations in record time.

At 7am on Saturday 8 May 1943 the *Victorious* finally slipped her moorings and left Pearl Harbor with the battleship USS *North Carolina* and the destroyers *Smith*, *Pringle* and *Case*, to join Task Force 14 of the United States Pacific Fleet. By 8pm on Tuesday 11 May the force was exercising in mid-Pacific in a position Lat 01° - 20'S/Long 167° - 26'W, having crossed the equator earlier in the day, but without any ceremony. The *Victorious* had on board 16 Avengers of 832 Squadron and 36 Martlets of 882, 896 and 898 Squadrons, and on the morning of Wednesday 12 May

Divisions on the flight deck during the Pacific sojourn of 1943. *(C. Minett)*

At Numea 1943, with USS *Saratoga* in the background. *(C. Minett)*

one Martlet crashed into the sea but the pilot, Sub-Lt Dixon, was rescued safely by the USS *Pringle*. Next day, as the force was off Fiji, they were joined by two more destroyers and at midnight on Friday 14 May they crossed the International Date Line and clocks were advanced by 24 hours, thus losing Saturday 15 May. During the afternoon of Monday 17 May the *Victorious* arrived at Numea on the French island of New Caledonia, which was the US Naval Base on their line of communication from Hawaii through Fiji and Numea, and which, like most of the French Pacific islands, had declared for General de Gaulle. However, the *Victorious* only stayed at the anchorage for 23 hours before, at 2pm on Tuesday 18 May, she set sail with other units of Task Force 14, including the battleships *North Carolina, Massachusetts* and *Indiana*, the cruiser *San Juan* and the aircraft carrier *Saratoga* which was flying the flag of Rear-Admiral De W. C. Ramsey USN. During joint exercises with the US carrier, the *Victorious*' personnel acquitted themselves well and they were able to equal the American carrier's impressive launching and landing rate. With the Japanese Fleet at sea and apparently heading for the Coral Sea, the task force was heading to intercept it, but by 20 May, it was clear that the Japanese ships had returned to their base and Task Force 14 turned back for Numea. Next day, whilst on passage, a Martlet crashed into the sea, but the pilot, Sub-Lt Wilkinson, was picked up safely by the battleship USS *Massachusetts*. The incident is recalled by John Herbert: 'Because of the extreme heat the air was relatively thin. Also our Martlet IVs were getting a bit teased out, and the net result was that we frequently sank over the bows on take-off and just skimmed the waves before crawling into the air. Clive Wilkinson, who subsequently got the DSC, didn't make it one time and went into the "oggin". As there were sharks around he got into his little rubber dinghy pretty smartly. We were then very surprised to see the flagship *Massachusetts* alter course and pick Clive up. Not only that, but the Admiral gave Clive his own cabin and also provided something to warm up the wet pilot. I can't imagine a British Admiral doing that, although to be fair Captain Mackintosh gave up his spacious cabin and his bathroom aft, to the young subbies who, because the ship was so crowded, were not even able to share an 8ft by 8ft cabin with two others. The Captain made do with his tiny sea cabin in the island. We slept in rough bunks which almost touched each other, but that was luxury compared with the mess decks.' Four days later, on the afternoon of Monday 25 May, the *Victorious* led the force back into the Numea anchorage.

During her stay at Numea Admiral Ramsey visited the ship to address the ship's company, and early on the morning of 1 June he hoisted his flag in the *Victorious* as she put to sea for three days of exercises. Also on board was Captain Mullinix USN of *Saratoga* and the American officers proved to be critical guests as they were shown how the Royal Navy operated. By the late afternoon of Thursday 3 June the *Victorious* had returned to Numea and next morning while launching three Martlets from the flight deck, one crashed into the sea, but both the pilot and the aircraft were recovered. Shortly after his visit Rear-Admiral Ramsey invited Captain Mackintosh on board the USS *Saratoga*, a carrier of some 33,000 tons which, together with her sister the USS *Lexington*, had been laid down in the early 1920s as a battle cruiser, but which had been completed in 1925 as an aircraft carrier, the world's largest at that time. It is interesting to read some of Captain Mackintosh's report on his visit: 'I visited the USS *Saratoga* at the invitation of Rear-Admiral D. C. Ramsey USN. I went to sea for three days of flying and gunnery exercises, taking with me 12 of my specialist officers. During this period I had the opportunity of studying American methods of operating a carrier and also comparing the design, construction and equipment of *Saratoga* with that of *Victorious*. All officers and personnel who have visited *Saratoga* have been most hospitably treated and given every assistance and information with regard to *Saratoga's* methods and equipment, which has done much to further the spirit of co-operation beween the two carriers, and to enable the *Victorious* to take her place in the task force commanded by Admiral Ramsey.

I was impressed by the fine spirit prevailing throughout the ship. The morale of the ship's company appeared to be of a high standard, this being more commendable when it is realized that there is a ship's company of approximately 3,200 men and that living quarters are therefore of necessity somewhat cramped, as in *Victorious*. (It is remarkable that this complement can be stowed in a ship the size of *Saratoga*). Also the ship has been based on New Caledonia for approximately eight months, where there is no night leave and where ordinary shore leave is very limited. Another marked feature of the ship was its very high standard of cleanliness throughout, as indeed is the case in all ships of the US Fleet that I visited. During our visit *Saratoga* landed her 68,000th aircraft which gives some idea of the operating experience of this ship.'

By mid-June the *Victorious* was a fully worked-up and integrated member of the US Pacific Fleet's Task Group 36.3 along with the USS *Saratoga*, USS *Massachusetts*, USS *Indiana*, USS *North Carolina*, USS *San Juan*, USS *San Diego* and USN destroyers *Mauny*, *Grindle*, *McCall*, *Craven*, *Fanning*, *Dunlop*, *Cummings*, *Case*, *Selfridge*, *Stanly*, *Claxton*, *Dyson* and *Converse*. On Wednesday 16 June, the whole task group put to sea in order to carry out intensive exercises in expelling massed air attacks, and aircraft from both the *Saratoga* and *Victorious* cross-operated, with six Tarpons and 12 Martlets being exchanged for eight Avengers, six Dauntlesses and 12 Wildcats. Although one Martlet crashed on deck there were few problems with landing signals now as *Victorious*' pilots had been using the US Navy's system since their arrival at Norfolk, Virginia, six months previously. Four days later, at 1.20pm on Sunday 20 June, the *Victorious* moored once again at Numea. On Tuesday 22 June a Martlet crashed into the sea near the ship, killing the pilot Lt-Cdr Metcalfe. Five days later the whole task group put to sea in order to provide air cover for the American landings on New Georgia in the Solomon Islands. This group of islands lies 200 miles north-west of Guadalcanal and extends for 150 miles in a north-westerly to south-easterly direction. The plan for the invasion envisaged landings at several points on New Georgia, with the capture of the main airfield being the main priority. On leaving Numea Task Group 36.3 steamed towards a position Lat 16°- 24'S/Long 159°-22'E, 'Point D', from where they would operate within a radius of 150 miles to cover the landing operations. They arrived in the early hours of Wednesday 30 June, ready for the main landing on New Georgia which took place at dawn, and by 7.30am almost all the US troops had gone ashore with very little opposition. For 28 days Task Group 36.3 remained at sea in the operating area with patrols being flown throughout each day and well into the night. On 1 July Avenger 4C crashed into sea, but the crew were rescued by USS *Fanning*. On Thursday 8 July a Martlet crashed into the barrier on landing and on Tuesday 13 July another crashed into the sea, with the pilot Sub-Lt Madden being rescued by the USS *Selfridge*. Three days later another Martlet crashed on deck, then on Wednesday 21 July, with the New Georgia landings having been completed successfully, Task Group 36.3 set course for Numea once again. Three days later, once the islands were within sight, the US Navy's Wildcat Squadron were flown off from *Victorious* and the Avengers of 832 Squadron returned. It had been the first occasion when a British squadron had operated from a US aircraft carrier. At just before noon on Sunday 25 July the task group anchored at Numea after their successful operation on New Georgia which was, in fact, a prelude to the capture of Bougainville, but this would be a far more difficult task. Throughout the period at sea the *Victorious* had operated as a fighter carrier, with 60 F4F variants embarked - 36 of her own Martlets and 24 F4F-4s of VF-3 from the *Saratoga*. The US carrier retained 12 Wildcats, primarily because the *Victorious* could take no more, and in addition to her own 36 Dauntlesses and 20 TBFs she had 16 TBFs of 832 Squadron embarked. A total of 614 sorties were flown by the *Victorious* during this long period at sea, and she was refuelled twice, by *Cimarron* and *Kaskaskia*. In addition to the 3,270 tons of furnace fuel oil and 30,000 gallons of aviation spirit, the *Victorious* received 20 gallons of ice cream and even 800lbs of dehydrated potato which arrived in the bomb-bay of a TBF, on which was printed the name *Spud Express*. Unfortunately, the dried potatoes were not always appreciated on board the *Victorious* and the US Navy could not help out with the rum shortage, which fortunately the Australian Navy was able to remedy.

For the *Victorious* 28 continuous days at sea had set a new record and at noon on the last day of July, with her ship's company manning the flight deck, and two Japanese prisoners of war embarked, she left Numea and the US Pacific Fleet. With the new Essex and Independence-class aircraft carriers about to be commissioned into the US Navy, the urgent need for a British carrier in the Pacific had been removed, but the US Navy was not going to let her departure go unnoticed and on board the *Saratoga* the band paraded to play her out with 'California Here We Come'. She was escorted by the USS *Indiana* and a destroyer screen which included the USS *Converse*, *Boyd* and *Halford*. During flying operations on the morning of Wednesday 4 August a Martlet was lost when it crashed into the sea, but, fortunately, the pilot was rescued by the USS *Boyd* and he was transferred back to the carrier that evening. Five days later, at 12.15pm on Monday 9 August, the *Victorious* arrived at Pearl Harbor where, during a three-day stay, the two Japanese prisoners were disembarked and some of the additional gun sponsons, which had been fitted earlier in the year, were removed.

It was at 8am on Thursday 12 August that the *Victorious* steamed out of Pearl Harbor to the strains of three US Navy Bands which turned out to play her out of harbour, one of which was in a launch which sailed alongside for as long as it could keep up. Six days after leaving Pearl Harbor the *Victorious* arrived at San Diego for a stay of just over 24 hours, which gave many members of the ship's company enough time to visit the Mexican border. On the morning of Thursday 26 August the Panama Canal pilot was embarked and by 7.15pm that evening she was berthed alongside Pier 6 at Cristobal where, as before, there was no blackout and the lights twinkled brightly. Two days later she left the Panamanian port bound for the US Naval Base at Norfolk, Virginia, where she arrived on Wednesday 1 September to find the *Indomitable* berthed and undergoing major repairs. After a period in dry dock when the gun sponsons were replaced following the Panama Canal passage, the *Victorious* left Norfolk at 8am on Thursday 16 September and ten days later she was back in the Firth of Clyde where all serviceable aircraft were flown off to HMS *Condor*, the Royal Naval Air Station near Dundee. The *Victorious* anchored for the afternoon in Rothesay Bay and at just before 5pm she weighed anchor and set course for the River Mersey. Next morning, at 11.40am, she was secured in the graving dock of Gladstone Dock, Liverpool, and as she came under refit routine the starboard watch departed for a well-earned long leave.

Although she had not seen action during her Pacific sojourn, she had acquitted herself well and had earned the praise of the US Navy's Admiral Chester W, Nimitz, the C-in-C Pacific Ocean Areas.

HMS *Victorious* at sea in the Pacific, whilst operating with USS *Saratoga* in 1943. *(Fleet Air Arm Museum)*

Chapter Six

'Operation Tungsten' And The *Tirpitz*

For the remainder of 1943 the *Victorious* lay in Liverpool's Gladstone Dock, not far from her sister ship *Illustrious*, which was undergoing modifications to her flight deck having provided cover for the Salerno landings in the Mediterranean. At 11.40am on Monday 1 November 1943, the lower deck was cleared and Captain L. D. Mackintosh said goodbye to the ship's company before handing over command temporarily to Commander R. C. V. Ross RN, who would be relieved on Wednesday 8 December 1943 by Captain M. M. Denny CB CBE RN. The son of a Gloucestershire vicar, Captain Denny had joined the Navy at 13 years of age in 1909 and had attended the RN colleges of Osborne and Dartmouth before going to sea as a midshipman in 1914. During the Great War he had served in the battleships *Neptune* and *Royal Sovereign*, seeing action at Jutland in the former ship. He specialized in gunnery in 1920 and spent a good deal of time in the inter-war years at HMS *Excellent* in Portsmouth in gunnery experimental work. Between 1937 and 1940 he had commanded the cruiser *Kenya*, before becoming Chief of Staff to the C-in-C Home Fleet and so his appointment to the *Victorious* was his first experience of an aircraft carrier and naval aviation.

The *Victorious* remained in Liverpool's Gladstone Dock until 12.15pm on Thursday 10 February 1944, when she set course downriver for the Irish Sea, and at just before midnight she arrived at Greenock. Next day, as daylight broke, it could be seen that the Clyde was crammed with all manner of merchant shipping, including the Cunard liner *Queen Elizabeth* as well as naval units. The latter included the battleships *Nelson* and *Ramillies*, the aircraft carrier *Argus* and at least nine smaller escort carriers. It had been a long time since the *Victorious* had been part of such a large British fleet and that morning she was visited by Admiral A. L. St G. Lyster, Rear-Admiral Aircraft Carriers, and the personnel of 829 and 831 Barracuda Squadrons joined the ship. Next day, before dawn, the *Victorious* weighed anchor and put to sea to embark the aircraft of both squadrons and to start her post-refit trials and work-up. However, machinery problems were experienced during this time and after only four days the carrier returned to Liverpool, on the afternoon of Wednesday 16 February, and tied up again alongside the North Wall of Liverpool's Gladstone Dock. Meanwhile, the Barracudas of 827 and 829, and the Corsairs of 1834 and 1836 Squadrons were preparing for the carrier's next big operation - against an old adversary.

During the *Victorious'* months with the US Pacific Fleet

Back in northern waters and more heavy seas.
(C. Minett)

55

the German battleship *Tirpitz* had continued to keep powerful British naval forces tied down, and in September 1943 an attack by midget submarines had succeeded in damaging the battleship. While the German Navy were carrying out repairs, the Soviet Air Force made an attack on her with 15 bombers, but they succeeded in inflicting only minor damage and by 15 March 1944 the *Tirpitz* was once again seaworthy and able to steam at 27 knots. Progress in the effecting of the repairs had been closely monitored by British Intelligence at Bletchley Park and the new C-in-C Home Fleet, Admiral Sir Bruce Fraser, was aware that this powerful battleship was due to carry out further sea trials in early April 1944. He decided to launch a Fleet Air Arm strike on the *Tirpitz*, either at sea or whilst she was at anchor in Kaa Fjord, in northern Norway, just south of Hammerfest. The force would include the *Victorious*, together with the escort carriers *Emperor*, *Fencer*, *Pursuer* and *Searcher*, as well as the rather elderly carrier HMS *Furious*. They were to be escorted by the battleships *Anson* and *Duke of York* (flying the flag of Admiral Fraser), the cruiser *Belfast*, which had taken part in the successful action against the *Scharnhorst* in December 1943, the *Jamaica* and the destroyers *Ursa*, *Swift*, *Undaunted*, *Milne*, *Nestor* and *Virago*.

For the operation the *Victorious* was to embark 21 Barracudas of 827 and 829 Squadrons and 28 Corsairs of 1834 and 1836 Squadrons. The US-built Corsairs were large, fast aircraft which had proved their worth both as fighters and as fighter-bombers, and it would be the first time that they had operated from the *Victorious*. All the aircraft had undergone training on a dummy range which had been built at Loch Eriboll on the north coast of Scotland, resembling the *Tirpitz's* anchorage in Kaa Fjord.

The *Victorious* finally left Liverpool during the morning of Tuesday 7 March, and that same evening she anchored in Rothesay Bay in preparation for starting her work-up next day. After embarking the squadron personnel her aircraft were flown on safely in the Firth of Clyde and on Thursday 9 March she started an intensive period of flying exercises in the Irish Sea. The fact that the *Victorious* was carrying more aircraft than she had been designed to operate soon began to cause problems when, on the morning of Friday 10 March, the last Corsair to land on crashed into another parked aircraft, damaging them both. Later that afternoon, at 4.50pm, a Corsair of 1834 Squadron crashed in the North Channel with the loss of its pilot, Sub-Lt P. C. O'Conner RNVR. This intensive training continued during most of March, with the ship anchoring in Bangor Bay or in the Clyde at the end of each day. Finally, on the morning of Friday 24 March the *Victorious* left the Clyde to steam north to Scapa Flow where she anchored during the afternoon of the next day.

On Monday 27 March a large convoy, JW58, consisting of 49 merchant ships, left Loch Ewe for Archangel and three days later the naval force sailed from Scapa Flow to start its passage towards northern Norway. Its mission was to provide cover for the largest convoy ever to sail for northern Russia, and to be prepared for the *Tirpitz* breaking out and mounting an attack on the convoy. By Sunday 2 April, with the force in a position Lat 70° - 47'N/Long 05°- 24'E, and rapidly approaching the launch point, Les Vancura recalls that Captain Denny announced to the ship's company that aircraft from the strike force would be attacking the *Tirpitz* whether she put to sea or not. Captain Denny was now faced with a number of problems, including the extreme cold and wet weather and the possible effect these conditions would have on the serviceability of the aircraft which had to be parked on the flight deck. Spray and sleet were freezing on the *Victorious'* flight deck and no one was really sure that the Barracudas and Corsairs, with their wings folded, could stand these conditions. In order to try to obviate any starting difficulties, the engines of all the exposed Corsairs were started and run up every two hours. This was no mean feat for the maintenance crews with the ship pitching heavily and with the flight deck slippery underfoot from frozen sleet and spray. Another concern for Captain Denny was the weekly shipping service between Trondheim and the North Cape, spotted by Corsairs which were on Combat Air Patrol in the vicinity of the naval force. Fortunately, the Corsairs managed to remain hidden in cloud and when it became clear that the force had not been observed, the *Trondheim Zenit* was allowed to proceed unmolested.

Meanwhile, on board the *Tirpitz* preparations were under way to get the battleship to sea for extended trials, which had been delayed because of bad weather. Kapitän Hans Meyer, the vessel's commanding officer, had ordered the removal of all the torpedo nets in preparation for departure during the forenoon of Monday 3 April and this intelligence caused the Navy's strike on the *Tirpitz* to be brought forward by 24 hours. This meant there was very little sleep for the squadrons' aircrews on the night of Sunday 2/Monday 3 April, for at 1.30am on Monday they were called for briefing and 20 minutes later the first Corsairs were launched on Combat Air Patrols. All the aircraft had been fuelled and armed the previous evening and by 2.15am the first range of ten Corsairs and 12 Barracudas, with an additional Corsair on an outrigger, were being brought up the after lift and at 4.50am, with all the aircraft manned, engines were started. Ten minutes later flying off was commenced and on a comparatively calm sea it was completed in exactly 14 minutes - nine seconds, which is a tribute to the excellent preparations of the ground crews. Simultaneously, the *Searcher* and *Pursuer* launched 20 Wildcats while ten Hellcats took off from the *Emperor*. Seafires from the *Furious* and Wildcats from the *Fencer* kept guard over the force while the strike aircraft formed up and set course for their target at 4.37am. On

Barracudas from the *Victorious* en route to Alten Fjord to attack the *Tirpitz*. (C. Minett)

board the *Tirpitz* the ship's company were weighing the second anchor when they received a report of enemy aircraft at a distance of 43 miles and the defensive smokescreens were activated. 'Operation Tungsten' had achieved almost complete surprise.

The story can be taken up here by Lt-Cdr R. Baker-Faulkner RN, the Wing Leader of N8 TBR, who stated in his report: 'The first strike took departure from the fleet one mile west of HMS *Victorious* at 4.39am on a track of 139 degrees. This track was maintained for nine minutes to allow all the escorting fighters to take up their correct positions. The weather was good, $1/10$ cloud cover over 1,000ft, visibility extreme.

At 4.57am the striking force began to climb as we were then approximately 25 miles from the coast. Two minutes later Loppen Island was identified fine on the port bow, and our landfall close to the westward of that island was established. We crossed the coast at 5.08am when an accurate plot was abandoned in favour of map reading. The route followed was No 2, passing close to the westward of the head of Lang Fjord and thence eastward down the valley to the target anchored in position 'D' in Kaa Fjord. No interference from flak was encountered until the striking force was within three miles of the target area. At approximately 5.25am the striking force was deployed and began its initial dive from 8,000 feet. At the same time it began to come under heavy but inaccurate fire from HA batteries at the head of Kaa Fjord and elsewhere. My aircraft dived towards the mountain close north-west of the target, I pulled up over the top and dived steeply towards the target itself from a height of approximately 4,000 feet.

About ten miles from the target I disposed 830 Squadron aircraft astern of 827 Squadron's 12 aircraft. I deployed the starboard half of 827 Squadron shortly afterwards according to wing synchronized tactics. Sighted target in position expected then dived to keep hill cover sending all Wildcats and Hellcats down to strafe-gun the target. I then lost sight of all aircraft and carried out a dive from stern to stem of target, releasing bombs at 1,200 feet. Fighters had shot up target very well and undoubtedly spoilt *Tirpitz* gunnery. Smokescreen started to operate as we arrived in sight but was too late to be effective for at least the first aircraft. Smoke all round the fjord. Twenty-one aircraft attacked in 60 seconds exactly. Visibility throughout was exceptional and weather in general could not have been better. It is clear that a considerable element of surprise was achieved. No enemy fighters were sighted, and the flak in general was spasmodic, erratic and inaccurate.'

In the words of Captain Denny, 'All aircraft (first strike) returned in flight formation with a unanimous broad grin.' There was no doubt that this was well justified for they had scored 11 direct hits, as well as a number of near misses, and they had reduced *Tirpitz's* upper decks to a shambles. By 6.42am all the first strike aircraft had safely landed on and refuelling had commenced in the hangar. However, things had not started quite so well for the aircraft of the second strike as they began ranging aircraft at 4.35am, when one of the 12 Barracudas would not start and there was a delay of some seven minutes whilst, with some difficulty, it was struck down the forward lift. Flying off commenced at 5.15am, but another Barracuda, after

Bombs hit the German battleship *Tirpitz* which was in the process of leaving harbour.
(Imperial War Museum)

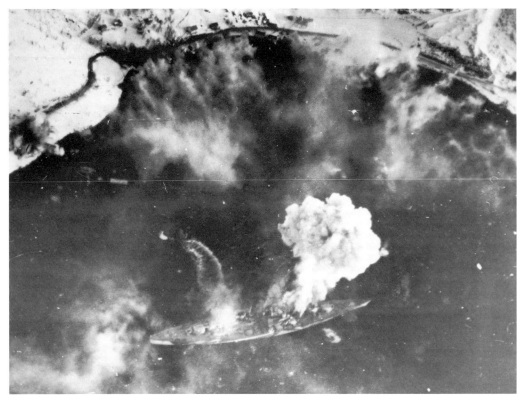

making an apparently successful take-off, developed a right-hand turn, which became steeper before the aircraft crashed into the sea with the loss of its three crew members, Sub-Lt Francis C. Bowles RNVR, pilot, Lt John P. Whittaker RNVR, observer, and Ldg Airman Colin J. Colwill, the Torpedo Air Gunner. By 5.34am the flying off was completed, which included the seven-minute delay whilst the non-starting Barracuda was struck down. In all there were 19 Barracuda aircraft and they were escorted by ten Corsairs, ten Hellcats and 19 Wildcats, and by 5.37am they had formed up and set course for their target. The story is picked up here by No 52 TBR Wing Leader Lt-Cdr V. Rance RN: 'A smoke float was dropped at intervals of a minute for three minutes after departure to give the escorting fighters some indication of the course taken by the second strike. The weather was fine, sea slight, swell nil. Cloud over target less than ¹/₁₀ at about 15,000 feet.

Initial approach from 10,000 feet was made with the TBR Wing in a shallow double "V" which, as no fighters were seen, was altered to two double columns when flak was encountered to facilitate evasion. The approach was continued so as to bring the target on the port bow, the starboard column being manoeuvred slightly back and up so as to keep the leader's column roughly between it and the target. Speed from crossing the coast was 165 knots for about 12 miles, then height was slowly reduced to increase the speed to 195 knots. After forming columns this was increased to about 210 knots and the final attack dive was commenced from between 7,500 and 7,000 feet.

Hellcats went down to attack heavy gun positions when these opened fire, and Wildcats attacked the *Tirpitz* herself immediately prior to the main attack. The final dive was carried out in quick succession, the port column diving first. The whole attack occupied about one minute. Medium dive-bombing except aircraft carrying 1,600lb AP bombs, which carried out steep glide bombing. Mean height of release and angle of dive were as follows: Aircraft carrying 1,600lb AP bomb - 3,000 feet - 45 degrees, remainder slightly under 3,000 feet - 50 to 60 degrees.

Considerable close-range flak mostly in the form of a box barrage round the target. *Tirpitz* had ceased firing by the time the last aircraft dived and fire with close-range weapons was opened much too early. One Barracuda was hit whilst over the target. It is believed that this aircraft carried out its attack in spite of this, but as it made its getaway with the remainder it was seen diving vertically onto the mountainside in flames. A large brown smokescreen had been laid from generators all round the target area and from the *Tirpitz* herself. This was visible 40 miles away. It did not interfere with bombing but must have hampered close-range weapons considerably. Unquestionably strafing attacks by fighters and the use of powerful blast bombs by the first few aircraft were of the utmost value in ensuring the safe arrival of the armour-piercing bombs carried by the latter half of the attacking forces. After clearing the coast on a course of 335°, two destroyers were first sighted and later the fleet itself.'

The Barracuda, LS 551, of 829 Squadron which had been shot down by anti-aircraft fire during the second strike was flown by Sub-Lt Hubert H. Richardson RNVR, with Sub-Lt Andrew G. Cannon RNVR as observer and Ldg Airman Ernest Carroll as Torpedo Air Gunner. As all the aircraft of the second strike began returning individually, with landing on commencing at 7.20am, it could be seen there had been no other casualties, although one Corsair failed to catch a wire and it sailed through the top of each barrier before crashing on its nose about 25 to 30 feet beyond the second barrier. Finally, one Barracuda which had not been able to jettison its bomb because of an electrical failure, was landed on last of all after the flight deck had been cleared of all personnel. Fortunately, the aircraft landed without a hitch, much to everyone's relief. The pilot of this aircraft was the late Lt-Cdr Dennis Phillips who later recalled the landing: 'As I approached I saw not a soul in sight except the batman contolling me with his bats. I smiled when he gave me the "cut" sign and immediately jumped into his safety net below the level of the flight deck.'

The attack had been a complete success and it was an example of what might have been achieved in 1941 had the *Victorious* carried fast, modern fighters and torpedo bombers, and had the same standard of intelligence been available then.

On board the *Tirpitz*, which had literally just cast off her mooring lines, the ship's company had little warning of the devastating strike which was only minutes away, and when the onslaught started they were still closing down the hatches and watertight doors. Kapitän Meyer had to rely on the surrounding smokescreen and the accuracy of the anti-aircraft batteries, both on board the ship and ashore, neither of which were particularly effective. Altogether the *Tirpitz* suffered 16 direct hits, with the worst devastation being caused in the boiler room air intakes, situated amidships. Two heavy bombs had caused some particularly serious damage with one of them hitting the ship just aft of the port catapult, smashing its way below before exploding on the armoured deck, demolishing bulkheads, ventilation ducts and power cables. One of the near misses from the first attack had exploded in the water alongside the ship, and it had penetrated under the armoured belt causing serious damage to an oil fuel tank. Altogether 122 members of the *Tirpitz's* ship's company were killed and 316, including Kapitän Meyer, were wounded, many of them by the devastating strafing fire from the escorting fighters who had done so much to disrupt the anti-aircraft barrage.

In his report to the C-in-C Home Fleet, Captain Denny praised his aircrews thus: 'I must place on record my unbounded admiration for the aircrews, both fighter and dive-bomber, who undertook this mission. The period of preparation and briefing prior to take-off was prolonged and arduous. A number of these young men are largely or totally unaccustomed to ship life. Few of them had ever engaged in an offensive operation before. It was impossible for them to get proper sleep the night before and they were called for briefing at 1.30am, that ghastly hour when man's stamina is at its lowest. Knowing that they were about to attempt an opposed attack of a character which hitherto had not been attempted by Allied aircraft in the European theatre, they left the carriers' decks in the greatest of heart and brim-full of determination, and proceeded through the complicated business of forming up and taking departure to the target exactly as if a parade ground movement. That in fact they met no German Air Force opposition subtracts not one iota from the credit that is due to them. As for the fighter escort, the following signal which was made from *Victorious* to the Rear-Admiral, Escort Carriers, indicates their magnificent performance: "Strike Leaders and aircrews of both strikes wish to thank the fighter escort for the superb way in which the fighters went down onto the flak."'

Originally it had been intended to carry out more attacks on the *Tirpitz* at dawn the following day, but Admiral Moore wisely cancelled these for a number of reasons, the main one being the fact that the battleship was obviously already seriously damaged. Another factor was the 'fatigue of the aircrews and their natural reaction after completing a dangerous operation successfully'. It was also clear to the Admiral that the enemy air defences would be fully alert, and in all likelihood they would have been reinforced, with the result that losses would be unlikely to be as low in a future raid. Instead he ordered the force to return to Scapa Flow where they arrived at just after 4pm on Thursday 6 April. The carrier force, led by the *Victorious*, formed line ahead as they approached the entrance to Scapa Flow and the event was graphically recorded by a press correspondent who had been on board the carrier since 30 March: 'The flagship and other ships of the Home Fleet cleared the lower decks, and gave three rousing cheers as the fleet carrier force that smashed the *Tirpitz* steamed line ahead into port. The carriers were welcomed home with full honours. As we passed each ship the officers and ratings lining the quarterdecks took off their caps and cheered, the sound reverberating across the blue waters. It was an inspiring sight, and, standing on the Admiral's bridge of the carrier, I felt very proud of being in such a ship. The sleek green Barracudas that had done the job were lined up astern on the flight deck and up forward, and the blue-uniformed pilots, observers and gunners stood in line. The Fleet Air Arm boys are happy tonight and they deserve to be.' Once the force had anchored safely in Scapa Flow messages of praise flooded in, including one from His Majesty King George VI which was relayed to the whole force and which read: 'Hearty congratulations on your

Captain Denny with His Majesty King George VI, during his visit to the ship on 11 May 1944. *(C. Minett)*

gallant and successful operation yesterday.'

Although 'Operation Tungsten' had been concluded, the *Victorious*' sojourn in northern waters was not yet over, and next day Captain Denny informed the ship's company that they had two further operations to carry out. What he did not say was that, once again, they would be attacks on *Tirpitz* in Alten Fjord. First of all, however, came a visit from Vice-Admiral Sir Henry Moore who addressed the ship's company on the morning of Friday 7 April, congratulating them on their magnificent effort just four days earlier. More exercises in the North Channel followed, with each night being spent at anchor in Bangor Bay or the Firth of Clyde until, at 10.30am on Tuesday 18 April, the *Victorious* anchored in Scapa Flow and rejoined the force which had accompanied her on 'Operation Tungsten'.

At just before 6am on Friday 21 April the *Victorious*, once again accompanied by the *Furious* and the escort carriers *Emperor*, *Pursuer*, *Searcher* and *Striker*, left Scapa for northern Norway, escorted by a cruiser and a destroyer screen. Originally it had been intended that another strike would be launched against the *Tirpitz*, but with atrocious weather conditions it was decided instead to attack shipping off the northern coast of Norway. However, with gale force winds and heavy seas, none of the carriers were able to fly off aircraft until the morning of Wednesday 26 April when a convoy of five ships was attacked, with three vessels being sunk and the other two damaged. Two days later the force was back at Scapa Flow where Les Vancura and friends were able to get ashore to enjoy an ENSA show. During the first week of May the *Victorious* carried out daily flying exercises, returning to her anchorage each evening, and on Monday 8 May General Sir Bernard Montgomery visited the ship and addressed the ship's company. Ray Barker remembers the occasion: 'The hangar was cleared and he stood on a partially elevated lift and addressed the ship's company. A small, wiry man with a slightly falsetto voice and a slight lisp, he thanked us for all we had done and were to do in the days to come. He spoke like a head of school, in prose peppered by expressions we had read in the *Hotspur*, *Rover* and *Wizard*. His dry delivery was not the stuff of famed orators, but his vitality, magnetism and self-confidence raised him to a new level.'

Three days later, at 11am on Thursday 11 May, the *Victorious* weighed anchor, stopping shortly afterwards to embark His Majesty King George VI, who came across from the destroyer *Searcher*. As soon as the King had embarked for his visit the Royal Standard was broken at the masthead and the *Victorious* put to sea where the royal visitor was treated to a flying and gunnery display. On the return to Scapa Flow at just after 6pm the King inspected the ship's company at Divisions before disembarking.

Commander Ross, the carrier's Executive Officer, recalled the visit: 'I had just 48 hours' notice of the King's visit. The ship was looking careworn and a little scruffy so I arranged for a painting party to commence work and they went to it with a will. Unhappily the provisioning party embarked a load of flour and peas over a wet paint area, with the result that it looked almost as bad as before. In the end I decided to paint only those parts of the ship that HM would see on his way from the starboard gangway to the bridge. Luckily his barge did not arrive on the port side. Whilst we were at sea, Captain Denny decided he could not leave the bridge so I had the privilege of escorting the King to lunch in the wardroom. We had a small table, Engineer Commander Cronk, Lt-Cdr Pollock, a Squadron Commander from the morning's flying, and myself. We had grapefruit, fried chicken and vegetables, tinned pears and rice (the King declined this dish), cheese and biscuits, and coffee. The King enjoyed a whisky and soda. Later that day we had tea with Admiral Fraser. The King, the Admiral and I (plus telescope) crammed into the tiny lift from the bridge to the wardroom flat six decks below. "May I press the button?" asked His Majesty. I graciously gave him permission. Wherever we went that day I had a lieutenant scouting ahead for the inevitable ordinary seaman carrying a bucket of dirty water. One was always lurking somewhere; they did it to Admiral Nimitz and they could do it to the King. When His Majesty left, Captain Denny - believing I had had my share of fun - banished me to the flight deck where I had to listen on the breeze for the Captain shouting, "Hip! Hip!", then I had to join in and ensure that all assembled provided a good, loud, resounding "HOORAY!" '

Next day, at 2.15pm, Vice-Admiral Sir Henry Moore hoisted his flag in the carrier and a few minutes later, together with the *Furious*, she was under way, escorted by the cruisers *Kent* and *Devonshire* and a screen of destroyers. Once again the force was bound for the Norwegian coast to carry out raids on shipping in the area, but inclement weather delayed flying operations until the afternoon of Monday 15 May when 27 Barracudas escorted by fighters were able to take off. However, the weather conditions soon deteriorated once again and Captain Denny announced that the force would be returning to Scapa Flow, where they arrived at just after 7pm on Thursday 18 May. Next morning, after refuelling, the *Victorious* left Scapa Flow and steamed south for Liverpool, and at just before noon on Saturday 20 May she secured alongside the North Wall of Gladstone Dock for a six-day maintenance period.

It was at just after 3pm on Friday 26 May that the *Victorious* slipped her moorings and left the Mersey to return to Scapa Flow where she arrived the next day. No sooner had she refuelled than Vice-Admiral Sir Henry Moore embarked and a few minutes later, at 1.15pm on Sunday 28 May, *Victorious*, *Furious* and the cruisers *Berwick* and *Devonshire*, together with a destroyer screen, set sail once again bound for the Norwegian coast. As in the previous sortie they had orders to attack shipping in the area, and RAF reconnaissance aircraft had reported a small convoy between Bergen and Trondheim which, Captain Denny announced, was their target. By 8pm on Thursday 1 June the force was in a position Lat 62° - 59'N/Long 03° - 33'E, and with the convoy having been sighted off Stadtlandet, a strike force of 16 Barracudas, escorted by Corsairs from the *Victorious* and Seafires from the *Furious*, was launched at 8.37pm. The convoy which had been sighted off the promontory of Stadtlandet consisted of the supply ships *Leonhardt* (4,000 tons), *Florida* (5,500 tons), and the *Sperrbrecher 181* (2,300 tons), escorted by destroyers and anti-aircraft flak ships. While the Corsairs and Seafires strafed the flak ships, the Barracudas attacked the supply ships, sinking the *Leonhardt* and the *Sperrbrecher 181* and badly damaging the *Florida* which was subsequently beached. Unfortunately, one Seafire from the *Furious* was lost and one Corsair from the *Victorious* failed to return with the loss of the pilot, Sub-Lt Ball RNVR. By 10.15pm all the remaining aircraft had landed on safely and 24 hours later the *Victorious* anchored in Scapa Flow.

Five days later, in the early hours of Wednesday 7 June, flying the flag of Rear-Admiral McGrigor, Admiral Commanding 1st Cruiser Squadron, the *Victorious* left Scapa Flow with the *Furious* and a large force of cruisers and destroyers on a diversionary operation to try to distract some attention from the Normandy landings which had started the previous day. The force sailed close to the Norwegian coast, but encountered no enemy forces and by 10.15pm on Thursday 8 June they had returned to the anchorage at Scapa Flow.

The *Victorious* had had a busy but successful three months in northern waters, beginning with the assault on the *Tirpitz* and following this up with attacks on enemy shipping off Norway. The German coastal traffic was of great importance to their hold on the north of that country, for it was the only means of communication with the garrisons there, and the Fleet Air Arm attacks had seriously disrupted the resupply of these outposts. In addition to this the new American-built fighters had shown that they were more than a match for the ME 110s and FW 190s.

However, it was to be the *Victorious*' last wartime foray into the cold northern waters of the Norwegian Sea and Arctic Ocean and she was now bound for warmer waters as promised by Captain Denny.

Chapter Seven

East of Suez

During the evening of Friday 9 June 1944 the *Victorious* left Scapa Flow for Greenock, together with the *Indomitable* which had only recently returned from a long refit in the USA following torpedo damage sustained in the Mediterranean in July 1943, and both ships anchored there the following day. However, the stay in the Clyde lasted for less than 48 hours and at 1.40pm on Monday 12 June the two carriers sailed for Algiers, in company with the destroyers *Undaunted*, *Undine*, *Urchin* and *Vigilant*. Embarked were 1834, 1836 and 1838 Squadrons of Corsairs, together with the Barracudas of 831 Squadron. As they approached the Strait of Gibraltar, with the weather being sunny and warm, the ship's company changed into tropical rig. It was over 18 months since the *Victorious* had last been in the Mediterranean and then it had been a very hostile place during the Allied landings in North Africa, but now conditions were very different and when the ships tied up alongside the main mole of Algiers Harbour shore leave was piped and hundreds of matelots made for the local Casbah. Next day, at just after 4pm, the carriers left Algiers and set course for Alexandria. During the passage routine flying exercises took place and on the afternoon of Tuesday 20 June a Corsair crashed into the sea, but, fortunately, the pilot was picked up safely by HMS *Teazer*. However, on the following morning there were two tragic accidents which led to the loss of seven lives. The first accident happened at 11.48am when Barracudas LS 732 and LS 581 were involved in a mid-air collision astern of the *Victorious*. The ship was stopped immediately and the seaboat launched, but the bodies of only two crew members, Sub-Lt I. C. Foskett and Ldg Airman W. E. V. Watkins, were recovered.* That afternoon, at just after 4pm, Corsair JT 386 crashed into the sea on take-off with the loss of its pilot Sub-Lt M. C. Kelly. It was a very sad day, and at 7pm that evening the ship slowed right down as the bodies of the two crew members were committed to the deep and a memorial service was held to the memory of the five men whose bodies had not been recovered. At 6.45pm on Thursday 22 June the two carriers secured to buoys in Alexandria Harbour where, once again, leave was piped and many went off to enjoy the city, making use of local feluccas to get ashore. It was just after 10.30am on Saturday 24 June when the *Victorious* and *Indomitable* left Alexandria and at 9pm that evening they commenced their southbound transit of the Suez Canal, with the *Victorious* leading the way. That night, with a huge searchlight at the forward end of the flight deck and two others on the bridge wings to light up the canal banks, the two carriers made their way slowly south over the 195 kilometres of waterway, passing El Kantara at just after midnight and Le Deversoir at 5am. By 9.45am on Sunday morning the *Victorious* had anchored in Suez Bay to await the arrival of the *Indomitable* and an hour and a half later both carriers set course for Aden.

During their passage of the Red Sea temperatures on board soared to over 100°F and many members of the ship's company laid out their hammocks on the flight deck and weather decks, while down in the boiler and engine rooms the personnel on duty were allowed to work shorter watches. On the morning of Wednesday 28 June the *Victorious* and *Indomitable* arrived at Aden, where they stayed just long enough to be refuelled before leaving at 6am the next morning for Bombay. Once in the Arabian Sea the fresher south-westerly breezes made life on board much more comfortable for everyone and salt water showers in all the ship's bathrooms alleviated the fresh water situation on board. During the crossing of the Arabian Sea the aircraft from both carriers cross-operated, and at 11.30am on Friday 30 June there was an accident involving one of the *Victorious*' Barracudas, LS 637, which crashed into the sea after being launched from the *Indomitable*. The pilot and Torpedo Air Gunner survived the accident and they were rescued by the destroyer *Napier*, but the observer, Sub-Lt G. D. Patterson RNZNVR, lost his life and his body was recovered by the destroyer. At 2pm that afternoon, when the ships were in a position Lat 14° - 09'N/Long 52°- 03'E the funeral service was held for Sub-Lt Patterson and his body was committed to the deep.

The force arrived at Bombay on Monday 3 July, and after a short stay of just 48 hours they steamed south for Ceylon (Sri Lanka), arriving in Colombo Harbour at 2pm on Friday 7 July to join the Royal Navy's Eastern Fleet. For most of the ship's company the island of Ceylon seemed like a tropical paradise, with the highlight of any run ashore being the beach at Mount Lavinia with its coconut palms and Indian Ocean surf, but for over two years the island had been the Royal Navy's forward base for the war against Japan.

In December 1941 the Royal Navy's presence east of Suez had been wholly inadequate, and even the reinforcement of the fleet by the elderly battlecruiser *Repulse* and the much

---

* The four other aircrew who lost their lives were: Sub-Lt W. Smith; Sub-Lt G. E. Grindrod; Lt J. H. Lewis; Ldg Airman I. C. Kitley.

The *Victorious* makes her southbound transit of the Suez Canal on the morning of Sunday 25 June 1944. *(C. Minett)*

newer battleship *Prince of Wales* at the recently completed, and much vaunted, Naval Base on the east side of Singapore's Johore Strait, did nothing to deter a powerful and aggressive Japan which was only too well aware of Britain's weakness in South-East Asia. The Japanese invasion of Malaya at just after midnight on Monday 8 December, and their sinking of the two capital ships two days later gave some idea of the formidable enemy that the Royal Navy was up against in the Far East. The loss of the *Prince of Wales* and the *Repulse* was a disaster for the Royal Navy and as Winston Churchill wrote: 'In all the war I never received a more direct shock . . . As I turned over and twisted in bed the full horror of the news sank in upon me. There were no British or American capital ships in the Indian Ocean or the Pacific except the American survivors of Pearl Harbor, who were hastening back to California. Over all this vast expanse of waters Japan was supreme, and we everywhere were weak and naked.'

However, this was by no means the end of the catastrophes in the Far East, for on 15 February 1942 the British garrison at Singapore surrendered and 14 days later, in the closing stages of the Battles of the Java Sea, the cruisers *Exeter* and HMAS *Perth* were both sunk, together with the destroyers *Electra*, *Encounter* and *Jupiter*. By the end of March the Royal Navy had been forced back to its base at Trincomalee, and in early April 1942 Japanese naval aircraft sank the cruisers *Dorsetshire* and *Cornwall* west of Ceylon and four days later, on 9 April, they located and sank the elderly aircraft carrier *Hermes* off Batticaloa in Ceylon. Admiral Sir James Somerville, who had been appointed to command the British Eastern Fleet in March 1942, was forced to move his headquarters from Trincomalee to Kilindini, Mombasa, although, unbeknown to the West at the time, the Japanese Navy had, in fact, reached its operational limits. The Japanese naval force which steamed into the Indian Ocean included the aircraft carriers *Akagi* and *Kaga*, both of which had taken part in the Japanese attack on Pearl Harbor. However, within three months of this aggressive sortie into the Indian Ocean both of them were sunk by American dive-bombers at the Battle of Midway as the United States rapidly went on the offensive in the Pacific theatre. As for the Royal Navy, the remainder of 1942 and the whole of 1943 was a period of recuperation as the Eastern Fleet was slowly built up once again into an aggressive fighting force. It was not until the surrender of Italy in September 1943 that any strong reinforcements could be deployed to join the Eastern Fleet and later that month the escort carrier HMS *Battler* was sent to Bombay to carry out convoy protection duties. From late March 1942 until January 1943, when she returned to the UK for a six-month refit on the Mersey, the *Illustrious*, and for just six months in early 1942, the *Formidable*, were the only British aircraft carriers with the Eastern Fleet, and their operations had often been constrained by lack of escorts, so the arrival of the *Victorious* and *Indomitable* in July 1944 to join five escort carriers was a prelude to the Royal Navy taking the offensive.

On Sunday 9 July 1944 Admiral Somerville, the C-in-C Eastern Fleet, visited the *Victorious* where he inspected Divisions, and next day the ship's company started an eight-day self-maintenance period before, at 11am on Wednesday 19 July, the carrier left Colombo. Once at sea she embarked her squadrons and at 6.40pm that evening she moored in the picturesque anchorage of the Naval Base at Trincomalee,

63

joining the rest of the Eastern Fleet there. By this time the Eastern Fleet had been reinforced by the escort carriers *Ameer*, *Atheling* and *Begum* which, combined with the *Victorious*, *Indomitable* and *Illustrious*, provided a powerful strike force. Admiral Somerville also had five capital ships under his command; his flagship *Queen Elizabeth*, the *Valiant*, the *Howe*, the French battleship *Richelieu* and the battlecruiser *Renown* (sister to the ill-fated *Repulse*), eight cruisers and a powerful contingent of destroyers. With this force at his disposal it was not long before Admiral Somerville went into action, and at 3pm on Saturday 22 July he left Trincomalee flying his flag in the battleship *Queen Elizabeth*, together with the battlecruiser *Renown*, the *Valiant*, *Richelieu*, seven cruisers, among them the Dutch *Van Tromp*, and ten destroyers. All these ships, which were designated Force 62, were to provide the escort and bombardment force for the Fleet Air Arm strike force aboard the *Victorious* and *Illustrious* during 'Operation Crimson', the bombing and neutralization of airfields on the small island of Sabang at the northern tip of Sumatra. In addition the submarines *Templar* and *Tantalus* were employed on air-sea rescue duties and were stationed between the island and Force 62. In view of what was becoming known about the brutal treatment which the Japanese were meting out to Allied prisoners of war, and to captured aircrew personnel in particular, this was considered to be an important duty which would do a great deal to boost morale. During the night of 24/25 July the aircraft carriers detached from the main group and, screened by the *Phoebe*, *Roebuck* and *Raider*, they steamed to their operating area which was centred 35 miles west-north-west of Sabang, in a position Lat 06° - 01'N/Long 91° - 47'E. They arrived in the early hours of 25 July and meanwhile the battleship, cruiser and destroyer bombarding forces closed Sabang from the north to reach their bombarding positions by 6.40am. The first strike launch had been scheduled for 33 minutes before sunrise, but this proved to be far too early as the morning was unusually dark and flying off was delayed for another five minutes. However, it was still too early, with the form-up being slow and clumsy and departure was further delayed. The first sortie of 1838 Squadron made a bad landfall and on arrival over the target it was still too dark for accurate machine-gunning. The enemy was on the alert and anti-aircraft guns opened fire as soon as the aircraft came within range, nevertheless, they did strafe the airfield and the Barracudas bombed the port installations at Sabang. However, as Captain Denny stated in his report to the C-in-C: 'Intelligence was such as to make it impossible to obtain good targets without breaking the element of surprise, for at 400 knots with flak opposition a camouflaged aircraft in revetment was not a very conspicuous target.' One Corsair was shot down by the anti-aircraft fire, but, fortunately, the pilot ditched at sea and he was subsequently rescued.

Meanwhile, the battleships were bombarding the harbour installations and military barracks of Sabang, whilst the cruisers shelled the wireless station on Pulo We Island and carried out counter-battery measures, with the destroyers bombarding a radar station. Aircraft from the *Illustrious* undertook spotting duties and the destroyers *Quality*, *Quilliam*, *Quickmatch* and *Van Tromp* under Captain (D) 4th Destroyer Flotilla, Captain R. G. Onslow, closed Sabang Harbour at speed as soon as the fleet bombardment was over, and shelled the harbour installations at close range. However, it was found impossible to silence the enemy guns for more than short periods and all four destroyers were hit causing some casualties, but damage was not serious. Later that afternoon, during the fleet's withdrawal, two enemy reconnaissance aircraft which attempted to approach were intercepted and shot down. Some time later a group of nine Japanese Navy Zeros were engaged by 13 Corsairs which shot down two of the enemy and damaged two more. Later the pilots reported that the efficiency of the Japanese pilots was not equal to that of 1942, but perhaps it was the fast, tough and well-armed US-built aircraft which had proved more than a match for the enemy. That evening the force steamed back to Trincomalee where they arrived, after a very rough crossing of the Bay of Bengal, at just before 3pm on Thursday 27 July. Although the operation had caused no more than an inconvenience to the Japanese it had given the aircrews valuable experience of their new enemy.

At Trincomalee there was now time to relax and although the facilities were not as plentiful as those at Colombo, Les Vancura discovered very nice beaches at Dutch Bay and Orlando Cove. It was announced that Noel Coward, who was in India to entertain the troops, would be visiting the fleet at Trincomalee and that the *Victorious* had been chosen to stage his concert. The event took place on the last day of July and all the aircraft were cleared from the after end of the hangar, with the partially raised lift being used as the stage. Many men came over to the *Victorious* from other ships of the fleet and several ratings sat on the flight deck with their legs dangling over the edge of the lift well. One pilot who was in the audience recalls that Noel Coward presented his 'West End Club' style repertoire in his impeccable Oxford accent, dressed in a white tuxedo, which was not really appreciated by the great majority of those present who preferred Music Hall style jokes and songs. However, the second half was received with more enthusiasm, although it was probably not his most successful concert. Commander Ross recalled: 'I went to meet Noel at the airfield. I was taken aback when his first words were, "Where's your Captain?" I suppose he expected red carpet, guard and band. Altogether he was rather like a Prima Donna. On the other hand he had just completed an exhausting tour of the troops in India which culminated in his being involved in a very nasty car crash. The doctors advised him to return home but he decided to visit Trincomalee and the Navy. Maybe we should have

This photograph, taken on 16 September 1944, shows the *Victorious* with Corsairs ranged on the flight deck ready for the raid on Sigli. HM Ships *Howe* and *Indomitable* are astern. *(Imperial War Museum)*

made allowances for his rather temperamental display.'

On 2 August the *Illustrious* left Trincomalee for a two-month refit in Durban, and Admiral Somerville's flagship steamed round the coast to Colombo. Next evening an ENSA concert party visited the ship to perform 'Love in a Mist', and they received a rapturous ovation which Ray Barker attributed partly to the presence in the party of some very pretty young ladies. Sadly the next day there was a tragedy when Barracudas LS 571 and MD 678 collided in mid-air killing three members of the aircrew of the former machine, Sub-Lt P. J. Hunter, Sub-Lt P. Dicks and Ldg Airman F. Hunt. Four days later on 8 August the battleship *Howe* arrived on station, the *Richelieu* returned from Colombo and the *Valiant* was manoeuvred into Admiralty Floating Dock 23. In the early hours of the following morning those on watch, and anyone else still awake, heard what sounded like a massive explosion. It was not, as might have been feared, Japanese saboteurs, but the sound of the floating dock breaking in half and sinking, and almost taking the *Valiant* down with it. Although the battleship stayed afloat she eventually had to return to England for repairs and she took no further part in the war.

In mid-August Admiral Sir Bruce Fraser arrived in Ceylon to relieve Admiral Somerville as C-in-C Eastern Fleet and it must have been a source of great satisfaction to him to see the *Victorious* which had operated so successfully under his command off Norway. Plans had already been made for the carrier's next offensive strike. Under the code name 'Operation Banquet' it had been decided that the target would be a large cement works situated at Padang on the west coast of Sumatra, about halfway down the island near the port of Emmahaven. Prior to the Japanese occupation the town had been the Dutch administrative capital of Western Sumatra and the Indaroeng cement factory was one of the biggest in South-East Asia, providing the Japanese with vital materials, particularly for the construction of airfields.

The object of the attack was to pin down Japanese air and naval forces while General MacArthur was developing his operations against Hollandia in New Guinea. For this strike the *Victorious* was accompanied by the *Indomitable*, whose squadrons had only been in the ship for three weeks. They were supported by the battleship *Howe*, the cruisers *Ceylon* and *Kenya*, and the destroyers *Rotherham*, *Redoubt*, *Raider*, *Rapid* and *Rocket*. This time the fleet was designated Force 64, and they sailed from Trincomalee at 11am on Saturday 19 August. The strike was delayed for 24 hours to enable the submarine *Sea Rover* to take up her air-sea rescue station, and on 21 August the *Victorious* crossed the equator, again with no ceremony. On the following day the cruisers and destroyers refuelled from RFA *Easdale* and on the morning of Thursday 24 August the force was in position off the west coast of Sumatra. On board the *Victorious* 'Action Stations' was sounded at 5.30am and 15 minutes later the first strike force of ten Barracudas, each armed with 500lb bombs, escorted by 19 Corsairs, was launched. The second wave, which was launched at 7.10am, consisted of nine Barracudas from the *Indomitable* and three from the *Victorious*, again armed with 500lb bombs and escorted by 12 Corsairs from the *Victorious*. Although the force did not meet any fighter opposition, there was some light anti-aircraft fire which shot down a Corsair of 1834 Squadron, killing the pilot Sub-Lt Thomas A. Cutler. The accuracy of the bombing was good with a lot of direct hits on the cement works, and pilots leaving the area reported a huge pall of black smoke hanging over the area, with tongues of flame shooting high into the air. They also reported clouds of cement dust in the air which obscured their vision as they set course for the *Victorious*. Aircraft which attacked the port of Emmahaven damaged two small ships which were alongside, but it was clear that the Japanese Navy had never made any use of the port.

Next afternoon, at 3pm, as the force was steaming back to Trincomalee, and while both the *Indomitable* and *Victorious* were carrying out flying operations, the *Victorious* suddenly hoisted 'Not Under Control' balls at the masthead as her rudder jammed. It was the first indication of a problem which was to stay with her for the rest of her career. After half an hour the *Victorious* was able to resume her station with the fleet and shortly afterwards she crossed the equator once again. At 12.15pm on Sunday 27 August she moored in Trincomalee Harbour for only three hours before steaming round to Colombo Harbour and flying all the serviceable aircraft ashore. As always Colombo and all its amenities was very popular with the ship's company and everyone enjoyed the break. However, the interlude was soon over and at 5pm on Thursday 14 September the *Victorious* and *Indomitable*, accompanied by the *Howe*, *Cumberland*, *Kenya* and seven destroyers, left Trincomalee to carry out an air strike on a railway repair and maintenance centre at Sigli on the northern tip of Sumatra. This was to have been preceded by a fighter sweep over airfields in the Medan and Belawan Deli areas, but it was cancelled on account of bad weather and instead aircraft from the *Indomitable* carried out photographic reconnaissance over northern Sumatra. At 6am on 18 September, in a position Lat 05°-45'N/Long 93°-55'E, ten Barracudas from the *Indomitable* and 13 Corsairs from the *Victorious* were launched. Unfortunately, one Barracuda from the *Indomitable* crashed into the sea, and once over the target area the attacking aircraft found heavy rain, shifting winds and very poor visibility, but they encountered no air opposition and very little flak. The force attacked at heights of between 2,000 and 3,000 feet and the standard of bombing was described as 'disappointing'. They completed the bombing in 60 seconds which was scarcely sufficient time for accurate aiming and, although the main targets

On 30 September 1944 the *Victorious* entered the Alexandra Dock, Bombay, for dry docking. (*C. Minett*)

were hit, some small targets escaped. Numerous mistakes and imperfections were apparent and the events of the disappointing day were compounded when two of the *Indomitable's* aircraft fired at HMS *Spirit*, the air-sea rescue submarine, while she was engaged in rescuing the crew of a crashed Barracuda. Fortunately, there was no damage or any casualties. The *Victorious* did not escape without some embarrassment either, for her Corsair Combat Air Patrol had been sent up without long-range fuel tanks and they had to be recalled whilst engaged in intercepting an enemy aircraft which had been reported by HMS *Howe*.

The mission ended at midday on Wednesday 20 September when the force returned to Trincomalee. In the Fleet Air Arm stores Les Vancura was getting an indication of the ship's long-term future when he was instructed to start exchanging all the Barracuda spares for Avenger parts, and he noted in his diary, 'This probably means that we are going to stay in the Pacific for a while.' However, the immediate future meant a maintenance period with dry docking and at 6.13pm on Friday 29 September, having flown off all the aircraft on leaving Ceylon, the *Victorious* anchored two miles outside Bombay Harbour. Next morning she steamed into the port's Alexandra dry dock and by that evening she was high and dry. It provided a welcome change of scenery for the ship's company, but by 2.15pm on Thursday 5 October the carrier was again anchored off the city and next day she set course for Colombo where she arrived two days later. Three days later, at 11am on Wednesday 11 October, she left Colombo for Trincomalee and during the passage she landed on the Corsairs of 1834 and 1836 Squadrons. Next morning, before entering harbour, she was subjected to mock air attacks by RAF Spitfires, but the Corsairs turned out to be more than a match for even the RAF's best fighters.

It was not long before the *Victorious* was in action once again and this time she and the *Indomitable* were to spearhead 'Operation Millet', attacks on the Nicobar Islands. Until this operation the attacks made by the Eastern Fleet had all been peripheral and of no great importance strategically, and there is evidence to show that the Japanese were aware that the Eastern Fleet was not strong enough to carry out any large-scale offensives and therefore had no intention of diverting any resources from their defences in the Pacific where the Americans were steadily pushing them back towards their homeland. However, the Nicobar Islands were different as they had moved troops there and were preparing the islands as strategic defensive bases to defend the approaches to the Strait of Malacca and so prevent the recapture of Malaya and Singapore.

The *Victorious* and *Indomitable*, together with the *Renown*, *London*, *Cumberland*, *Suffolk*, *Phoebe* and a screen of 11 destroyers, left Trincomalee at just after 7am on Sunday 15 October and set course for the Nicobar Islands. At 6am on Tuesday 17 October, when they were 30 miles east of Batti Malv Island, which is south of Car Nicobar, the two carriers parted company from the main force on a course of 45° for their flying off positions. Just over half an hour later, at 6.34am, the *Indomitable* began launching ten Barracudas all armed with 500lb bombs, and eight Hellcats, for a strike on shipping and installations in Nancowry Harbour. Four minutes later the *Victorious* flew off 19 Corsairs to attack airfields and eight to provide cover. The aircraft were soon over their target area and so complete was the surprise attained that the enemy defences did not begin firing at the aircraft for several minutes. However, once the anti-aircraft fire started it became a real problem for the *Indomitable's* aircraft below 3,000 feet. Unfortunately, one of her Barracudas failed to pull out of its dive and crashed, and two Corsairs from the *Victorious* crashed into the sea just north of Malacca with the loss of their pilots, Sub-Lts J. O. Charden and E. Hill. Nevertheless, despite these losses the bombing was accurate, with all bombs falling in the target area, and the 830-ton coaster *Ishikari Maru*, the only sizeable ship in the harbour, was sunk, as well as some smaller vessels. Later that day, at just after 5pm, the *Victorious'* rudder jammed again, which necessitated an emergency switch to steam steering.

That night, instead of withdrawing from the area, the force made a feint to the northward and Rear-Admiral Walker in the cruiser *London* was detached with two destroyers to bombard Car Nicobar before returning to Trincomalee. An air strike on Sabang had been planned for the next day, but as weather conditions were unsuitable, with torrential monsoon rains, the heavy units were ordered to bombard Car Nicobar. The carriers, meanwhile, were ordered to make further attacks on Nancowry and the *Victorious* should have launched 14 Corsairs at 7.15am, but just as they were all ranged and ready the carrier's rudder jammed and it was 7.45am before she was able to launch the aircraft. Although they caused little damage and four of them were hit by flak, all returned safely to the carrier. However, unlike the previous day's operations, at 8.40am a Japanese reconnaissance aircraft flew over the carriers, avoiding attempts to shoot it down, and soon afterwards at 9.30am a group of enemy aircraft were reported on radar. The raiders were missed by the *Indomitable's* fighters which were sent to intercept them, but they were later caught by fighters from the *Victorious* when they were 12 miles northeast of the carrier force, which at that time was ten miles south of Car Nicobar. The enemy force consisted of nine Japanese Army fighters (Oscars), and after a fierce mêlée lasting 40 minutes, during which covering aircraft from the bombardment took part, four enemy aircraft were shot down and others were damaged, with the loss of two Corsairs and one Hellcat. Meanwhile, the *Indomitable's* fighters, which had been recalled, were sent to chase three Oscars which had

High and dry in the Alexandra Dry Dock, Bombay. *(C. Minett)*

provided top cover for the enemy raiding force and which had turned away to the north-east. They were intercepted some 20 miles away and all of them were shot down without loss. The two days of action and the knowledge that the Fleet Air Arm's fighters could shoot down the best Japanese fighters had an excellent effect on morale.

Next day, as the force steamed back to Trincomalee, one of the escorting destroyers hit a mine and had to continue her passage at reduced speed, but the *Victorious* arrived back in the natural harbour at 8.30am on Saturday 21 October. On the following day she steamed round the coast to Colombo Harbour where stores and ammunition were embarked. By the end of the month the *Victorious* was back at Trincomalee but, as Les Vancura has recorded in his diary, 'We have a five-day paint ship and then, the "buzz" goes, we go to Bombay for a survey of the rudder.' On Thursday 9 November the *Victorious* put to sea for the day to carry out deck landing trials, and three days later the 'buzz' became reality when she left Trincomalee bound for Bombay where she arrived on 15 November. During the voyage news came from the UK that her old adversary, the German battleship *Tirpitz*, had finally been sunk at her moorings by the RAF, but on board the *Victorious* those cold northern waters seemed more than half the world away. Next morning she was manoeuvred into the Alexandra dry dock and by lunchtime she was high and dry once more.

This time the carrier remained in the dry dock for almost four weeks, during which time the rudder was removed and the steering gear was thoroughly overhauled. Each watch of the ship's company was granted four days' station leave at Deolali Camp some 100 miles north-east of Bombay. Ray Barker remembers his days there: 'In the morning the most welcome cry was that of the "Char Wallah" - the tea man - with a huge urn on his back, filled with strong, steaming hot tea, wonderful. That first morning cup in the camp remains one of my abiding wartime memories, a tropical nectar comparable to our Arctic kai. The camp boasted a concert party, quite as bad as that which I used to watch on Redcar Sands as a schoolboy in the late 1920s. The lack of talent was compensated for by sheer enthusiasm and hilarious spontaniety. Everyone slept under canvas and under mosquito nets. Wildlife was present in the form of various snakes as well as scorpions, one of which I found in my shoe one morning. There was no strict regime at the camp, with everyone being allowed to get up when they wanted and to come and go as they pleased. A small town nearby attracted many with its restaurants, shops, street vendors and the inevitable bazaar of stalls.'

There was one tragic accident during the stay in Bombay, when Able Seaman R. W. Richardson, who was carrying out sentry duty by the gangway on the edge of the jetty, fainted and fell into the dry dock. Although he was rushed to hospital, sadly he died later and his funeral was held next day.

On the morning of Thursday 14 December the *Victorious* left Bombay bound for Colombo where the *Indefatigable* was also moored, having left Portsmouth on 19 November. There were now four fleet carriers east of Suez and the stage was set for them to sail into the Pacific Ocean for the formation of the British Pacific Fleet.

Chapter Eight

The 'Meridian' Operations

With the end of the European War in sight the British Chiefs of Staff were keen to restore Britain to a position of genuine partnership alongside the United States in the Pacific, but the difficulty was in finding a credible way to do it. Prime Minister Winston Churchill had approached President Roosevelt with an offer of help in the summer of 1944, and by the end of the year a new British Pacific Fleet was more or less established at Trincomalee. The Second World War was to be the last major war in which Great Britain took part as a world power, and certainly as a leading protagonist in Asia. The speed of the country's withdrawal from the area in the years which followed the end of the war was an inevitable consequence of the catastrophic defeat at Singapore in February 1942 and of the economic exhaustion brought about by six years of total war. The operations in the Indian Ocean during 1944 had not been particularly rewarding for the Royal Navy, although 'Operation Millet' involved the carrier aircraft employing a two-phase attack for the first time in the war and the Eastern Fleet had carried out four offensive operations involving two fleet carriers on each occasion, which almost doubled the number of such operations executed by the Royal Navy since 1939. There were some other points of interest, one of these being the fact that the battleship *Queen Elizabeth* had engaged an enemy with her main armament for the

On Thursday 11 January 1945, the Supreme Allied Commander, South-East Asia, Lord Louis Mountbatten, visited the ship. Here he is seen on the flight deck with Captain Denny. *(C. Minett)*

first time since 1915, when she had fought off Gallipoli.

Unfortunately, amongst the British planners there were major differences over priorities. There were those who thought the recovery of the lost colonies of Malaya and, above all, Singapore should receive the highest priority while others considered that Britain should be involved in the Pacific War alongside the United States, which it was believed would give the country more influence at the subsequent political negotiations. In the event the latter thinking won the day and on 22 November 1944 Admiral Sir Bruce Fraser was ordered to form the British Pacific Fleet whose base was to be at Sydney. However, there was no prospect of the fleet leaving Trincomalee before the end of 1944 for it was well below its planned strength and, in addition, suffered a series of frustrating delays in re-equipping the carrier air groups with the American-built Grumman Avenger, which had been designed as a torpedo bomber but which had proved equally efficient as a fighter-bomber.

Many people have referred to Admiral Fraser's new command as 'The Forgotten Fleet', which is surprising for with no fewer than five fleet, four light fleet and seven escort carriers, four fast battleships, eight light cruisers, 28 destroyers, 33 escorts, and an auxiliary anti-aircraft ship, three fast minelayers and 22 submarines on station at the time of the Japanese surrender, the British Pacific Fleet was the most powerful single strike force assembled by Britain in the course of the Second World War. Admiral Fraser's new command gave him the delicate task of reconciling his responsibilities to the Admiralty for the safety and handling of his fleet and its supply train, not only with the demands pressed upon him by the Australian and New Zealand authorities (who would prove to be very difficult), but also with the subordination of his force to operational orders given to him by the Americans. Initially General MacArthur and Admiral Nimitz argued over who should be placed in overall operational command of the British Pacific Fleet, with MacArthur wanting to use it in support of the reconquest of the Philippines - which was still under way - while Nimitz thought it should be employed in support of the planned invasion of Okinawa. In the event Nimitz had his way and he and Admiral 'Bull' Halsey, two successive Commanders-in-Chief of the United States Pacific Fleet, employed Admiral Fraser's force as a flexible, self-contained reserve.

On formation of the British Pacific Fleet with Admiral Sir Bruce Fraser as its C-in-C, the Eastern Fleet was dissolved and Vice-Admiral Sir Arthur Power became the C-in-C of the newly designated East Indies Station. The ships which were to form the Pacific Fleet were the two battleships, *King George V* (Vice-Admiral Sir Bernard Rawlings, Second in Command BPF), and *Howe*; the four fleet aircraft carriers, *Indefatigable* (Rear-Admiral Sir Philip Vian, Flag Officer Commanding Aircraft Carriers BPF), *Illustrious*, *Indomitable*, *Victorious*; five cruisers, *Swiftsure*, *Gambia*, *Black Prince*, *Argonaut* and *Euryalus*, and three flotillas of destroyers.

Before leaving Trincomalee for Australia Admiral Fraser, in furtherance of the policy which had been decided upon by the Chiefs of Staff, of destroying Japanese war potential, particularly oil supplies, arranged that the C-in-C East Indies should carry out with the ships designated for the British Pacific Fleet a series of operations designed to destroy the main Japanese source of this commodity. It was known that the Japanese were short of oil and aviation fuel, for the American advances in the Pacific had virtually put the Japanese homeland under a state of siege, but oil supplies were still getting to Singapore. Admiral Power planned to put out of use the most important of the Japanese oil refineries at Palembang in southern Sumatra, by attacking them with maximum carrier forces as the ships of the British Pacific Fleet left the Indian Ocean on passage to Sydney. Although the Australian Government had been assured that the new fleet would reach Sydney before the end of 1944, because of the delays in receiving the new Avengers it was clear that this date could not be kept, so the attack on Palembang made sense and in the intervening weeks training exercises would be carried out.

The *Victorious* left Trincomalee on the morning of Tuesday 19 December 1944, to carry out gunnery exercises and to embark 21 Avengers of 849 Squadron. The ship went to 'Flying Stations' at just after 2pm and ten of the new aircraft landed on safely. However, at 2.50 pm one of the Avengers suddenly swerved to port as the pilot approached the flight deck and as the aircraft ditched over the side it hit and killed the Deck Landing Control Officer Lt K. F. E. Dorman. Fortunately, the aircrew were rescued safely, but that evening there was another funeral as the body of Lt Dorman was committed to the deep. Two more days of exercises followed before, on the evening of Saturday 23 December, the carrier moored in Trincomalee Harbour for a four-day Christmas break. Les Vancura has his own memories of that Christmas which, for most of the ship's company, was to be their last on board the *Victorious*: 'We were at anchor in company with the "*Indom*", *Illustrious*, *Indefatigable*, *King George V* and *Queen Elizabeth*, with the accompanying cruisers and destroyers. It was Christmas Eve, a lovely warm night, and we had been swimming over the side during the day. As we strolled around the flight deck everyone was talking about what their folks at home would be doing, and we were all terribly homesick. Suddenly across the anchorage from the loudspeakers of the flagship, *King George V*, came the sound of Bing Crosby singing "Silent Night". There was not a dry eye on the ship, nor I guess in the anchorage. We had formed a choir on the ship, taking part in ship's concerts and also in the carol service. On Christmas Day we went round the anchorage in a motor boat singing

The 'Meridian' raids on the Japanese oil refineries at Palembang in Sumatra were considered to be the most successful air raids mounted by British forces in South-East Asia. This series of photographs shows the damage inflicted by the Fleet Air Arm. *(C. Minett)*

carols, and we visited every ship in the fleet. When we got to the flagship, HMS *King George V*, we climbed aboard and threw three or four officers overboard from the quarterdeck - making sure first that none of them was the Admiral. They took it all in good part and turned their firemain hoses on us in retaliation as we pulled away.'

By 7.30am on Wednesday 27 December the Christmas festivities were well and truly over as the *Victorious* put to sea once again for gunnery and flying exercises, and this was repeated over the next two days when she operated with the *Indomitable* and *Illustrious*. New Year was celebrated at Trincomalee and at 10.15am on New Year's Day she sailed in company with *Indomitable* (Admiral Vian), *Illustrious* and *Indefatigable* for 'Operation Lentil'. This was to be the second rehearsal (the first having been carried out in mid-December by the *Indomitable* and *Illustrious*), for the attacks on Palembang. It had been decided that the carriers would bomb the oil refinery at Pangkalan Brandan (Pangkalanberandan), situated on the north-east coast of Sumatra, and Force 65, as it had been designated, would include an escort of the cruisers *Argonaut*, *Black Prince*, *Ceylon* and *Suffolk* together with eight destroyers.

The force arrived at the flying-off position north-east of Simular Island, off the west coast of Sumatra, in the early hours of Thursday 4 January 1945 and at just after 6am a fighter sweep of eight Hellcats from the *Indomitable* and eight Corsairs from the *Victorious* was flown off to attack airfields near the refinery. Although the operation entailed a difficult flight across the mountains of Sumatra, because there was a full moon it was considered wise to keep Force 65 inside the Malacca Strait, and with fine weather this plan would allow the whole force to remain in the area for much longer. The main strike of 16 Avengers from the *Victorious* and *Indomitable*, each armed with four 500lb bombs, took off at 7.40am, and they were escorted by 16 Hellcats from the *Indomitable*, 16 Corsairs from *Victorious* and 12 Fireflies from the *Indefatigable*. No opposition was met on passage to the target area, but on the final approach to the port of Pangkalan Seosoe they did encounter heavy and accurate anti-aircraft fire. Fortunately, over the target area the flak was much lighter and proved inaccurate and the Fireflies were detached to carry out rocket strikes. Next came the Avengers and although the target was obscured by flame, the bombing was accurate and considerable damage was done as oil tanks were set on fire. A force of enemy fighters (Oscars) east of the target area were engaged by the escort and five were shot down. One Avenger was damaged by an enemy fighter, and one had to ditch about 12 miles offshore with engine failure, but the crew were rescued safely. One Firefly ran out of fuel on the return journey, but, fortunately, he was within sight of the carriers and the pilot ditched in the water close to the *Indefatigable*. Meanwhile, the fighter sweep attacked Medan airfield and they reported destroying seven of the 25 enemy aircraft which were found on the ground, and shooting down two others. Belawan Deli was attacked and whilst no aircraft were lost, it was found that too many of the escorting fighters had been detached to strafe ground targets while four more chased and destroyed an enemy fighter, which left the bombers with very little cover as they made for their rendezvous after the raid. By 10.30am the force was heading back for Trincomalee, where the *Victorious* moored at 9.30am on Sunday 7 January.

During the following few days stores and ammunition were embarked and, on the morning of Thursday 11 January, Lord Louis Mountbatten, the Supreme Allied Commander, South-East Asia, visited the ship and spoke to the ship's company. He told them that they would soon be leaving for Australia to take part in the Pacific War, but first of all the *Victorious* put to sea that afternoon for flying exercises and to practise refuelling at sea. During Friday 12 and Saturday 13 January the whole fleet put to sea for two days of exercises which involved a full-scale rehearsal of the forthcoming attacks on the refineries at Palembang. Finally, at 3pm on Tuesday 16 January the British Pacific Fleet, designatd Force 63, left Trincomalee in very poor weather conditions to carry out 'Operation Meridian I', the destruction of the Pladjoe and Soengei Gerong oil refineries at Palembang. Following this the fleet would steam on to Australia, and the *Victorious* would not return to Trincomalee until after the war.

Altogether Force 63 carried a formidable complement of 238 aircraft as follows:

HMS *Victorious*: 849 Squadron-19 Avengers; 1834 and 1836 Squadrons-34 Corsairs.

HMS *Indomitable*: 857 Squadron-21 Avengers; 1839 and 1844 Squadrons-29 Hellcats, together with four Avengers and six Corsairs as spares for *Victorious* and *Illustrious*.

HMS *Illustrious*: 854 Squadron-18 Avengers; 1833 and 1839 Squadrons-32 Corsairs; 1700 Squadron-two Walrus (air-sea rescue aircraft).

HMS *Indefatigable*: 820 Squadron-21 Avengers; 887 and 894 Squadrons-40 Seafires; 1770 Squadron-12 Fireflies.

Force 63 met its oiling ships on 20 January and most of the day was spent fuelling in preparation for the approach to the flying-off position, between the west coast of Sumatra and the island of Engano, during the night of 21/22 January. However, owing to bad weather, Admiral Vian postponed the attack for 48 hours and it was the morning of Wednesday 24 January before conditions had improved sufficiently to commence flying operations. Initially it had been intended to make a heavy attack on Pladjoe refinery and also a photographic reconnaissance of the airfields and the main target, whilst escorting fighters raided the airstrip at the coastal town of Mana and other airfields in the area. But owing to the lack of information about enemy fighter strength, it was decided to neutralize enemy airfields at Palembang, Lembak and Talangbetoetoe as well. In the event ten Avengers from *Victorious*, ten from the *Indefatigable*, 12 from the *Illustrious* and 11 from the *Indomitable*, all armed with 500lb bombs, were to make the main strike at Pladjoe, escorted by 11 Corsairs from the *Victorious*, 11 Fireflies and 16 Corsairs from the *Illustrious* and 16 Hellcats from the *Indomitable*. The strike on the airstrip at Mana was to be carried out by four Avengers from the *Indomitable* and they were to be escorted by four Hellcats from the same carrier. The sweeps on the other three airfields were to be undertaken by 12 Corsairs from *Victorious* and 12 from *Illustrious*. However, in the event, four Avengers, two Fireflies and one Corsair were found to be unserviceable soon after take-off.

The main strike force began flying off at 6.15am that morning, and after having formed up they took departure at 7.04am with landfall being made 14 minutes later. Once over the coast all the aircraft began to climb over the Pegunungan Barisan mountain range which runs from north to south over the full length of Sumatra. Fortunately, with weather conditions good and a visibility of about 60 miles, the aircrew had a clear view of the scene and at a distance of about 20 miles, balloons were seen flying at 3,000 feet over the target. This was a most unusual defensive measure for the Japanese to employ and it indicated the importance which the Japanese attached to the Palembang oil refineries. The escorting Fireflies were instructed to shoot down the balloons, but transmission interference meant that they did not get the order. At 8.08am enemy anti-aircraft fire opened up on the bombers whilst they were still out of range, and then 25 Japanese Army fighters (Tojos) engaged the escort from above just as the bombers were deploying into circular formation. Fortunately, they remained unscathed and a few minutes later they began their attack, dropping their bombs from an average height of 3,500 feet with some of them being released from below the balloon barrage. As the Avengers completed their attacks they made for their departure rendezvous, 15 miles west of Palembang. At this point they ran into heavy anti-aircraft fire and they also encountered numerous enemy fighters which, it appeared, had been lying in wait for them. Unfortunately, owing to their flying off in the second range, the Corsairs which were making the airfield sweeps were unable to prevent enemy interference with the main strike, whose fighter cover was inadequate. The Corsairs and Hellcats were actually covering the rendezvous, with only the Fireflies patrolling the route from the target, the remaining fighters being engaged in air fighting south of the target. It was during this phase that several of the Avengers were badly damaged before the enemy, having lost at least 11 fighters with others damaged, disengaged at 8.25am. The bombers finally returned to their carriers at 10.23am.

Meanwhile, the Avengers from the *Indomitable* dropped 16 500lb bombs on the airstrip at Mana, and the fighter sweep destroyed at least 34 enemy aircraft on the ground as well as damaging numerous others, which prevented any sustained attacks on the fleet which was on station some 35 miles off the Sumatran coast. In all, six Corsairs, two Avengers and one Hellcat failed to return from the mission, plus a Corsair and a Seafire which ditched close to the fleet, but the pilots from these two aircraft were rescued safely. As soon as landing on was completed Force 63 steamed south-west at 22 knots, and although enemy aircraft were detected on the radar screen, they faded before the combat air patrols could engage them. There was no doubt that the attack had been a success and Japanese reports which were captured after the war showed that oil production at Pladjoe was halved as a result.

Force 63 refuelled on 26 and 27 January, and it was clear that the overall fuel situation would only allow for one more strike on Palembang. This time it was to be on the Soengei Gerong oil refinery, and designated 'Operation Meridian II', it was scheduled to take place on the morning of Monday 29 January. Although the fleet had maintained strict radio silence whilst off the coast of Sumatra, more fighter opposition was expected this time, and as a result of experience gained in the previous operation the fighter sweep was to fly in two parts, both of which were timed to arrive simultaneously over the two enemy airfields. On completion of the sweeps they were to establish patrols over the airfields and the route to the target area was altered to pass further south, in order to avoid the heavy concentration of anti-aircraft batteries which were situated north of Palembang. The standing air patrol which was to be provided by the *Indefatigable* would be backed up by 12 fighters from the other three carriers, for it was reasonable to infer that as only one of the two refineries at Palembang had been attacked, the Japanese would be anticipating a raid on Soengei Gerong and, therefore, an enemy attack on the fleet was a strong possibility.

The main bombing raid on the Soengei Gerong refinery was to be executed by 12 Avengers from the *Victorious*, ten from the *Indefatigable*, 12 from the *Illustrious* and 12 from the *Indomitable*, all again armed with four 500lb bombs. The escort was made up of 12 Corsairs from the *Victorious* and 12 from the *Illustrious*, together with nine Fireflies from the *Indefatigable* and 16 Hellcats from the *Indomitable*. The airfield sweeps were to be carried out by 13 Corsairs from the *Victorious* and 12 from the *Illustrious*, while the armed reconnaissance flights over Mana would be undertaken by two Fireflies from the *Indefatigable*.

The fleet arrived at the flying-off position at 6am on 29 January to find heavy rainstorms in a belt 30 miles off the coast, although visibility over the mountains appeared to be good. At 6.40am, when the carriers were in a clear patch between rainstorms, the main striking force began flying off, and with the form-up being somewhat prolonged they departed for their targets at 7.34am. The Fireflies were flown off in the second range, joining up shortly afterwards, and as soon as the main strike force was clear of the fleet the two airfield sweeps and the Mana reconnaissance flight took departure. The 13 Corsairs from the *Victorious* (Force Y) set course for Palembang, while the 12 Corsairs from *Illustrious* (Force X) headed for Lembak. One of the *Victorious*' aircraft ditched in the sea before reaching the coast and a Corsair from the *Illustrious* ditched shortly after take-off, with the pilot baling out. Meanwhile, in the main striking force, three Avengers were forced to ditch on their way to the target because of mechanical problems, but by 8.30am the rest were over the target. Unfortunately, the airfield sweeps found that most enemy aircraft were already airborne, although they did manage to destroy three Tojos at Lembak and one enemy aircraft was destroyed and two were damaged at Palembang.

Over the mountains it was found that the weather had deteriorated and the Corsairs which were providing top cover were forced to fly into clouds. However, once the target was sighted the Fireflies were sent to shoot down the barrage balloons which were flying at 4,000 feet, and almost immediately they ran into anti-aircraft fire. About five miles from the target enemy fighters attacked the rear of No 1 Wing (Avengers from *Victorious*' 849 and *Indomitable*'s 857 Squadrons), damaging one Avenger, nevertheless, at 8.45am the Avengers began their dive into the target from 7,500 feet, straight through the balloons, most of which were crashing down in flames. However, two of the aircraft were lost when they struck cables, but the bombing was more accurate than in the attack at Pladjoe, despite the fact that burning oil tanks again made it difficult for pilots to see the target. Once they had released their bombs the Avengers withdrew south of Palembang, then headed east to their rendezvous which was not easy to find. Some enemy air opposition was encountered at this stage and seven Avengers were surprised and damaged by up to nine enemy aircraft. However, the versatility of the Avenger was demonstrated when one of them followed an enemy Tojo down to a height of only 50 feet and shot it down. By 9.01am the main strike aircraft were able to leave the rendezvous and by 10.05am the *Victorious*' eight surviving Avengers had landed on. Six of the damaged Avengers had to ditch close to the carriers on their return, and on board one of these, from the *Victorious*, the observer, Sub-Lt Malcolm John Gunn of 849 Squadron, had been seriously wounded. In spite of the gallant efforts of the Telegraphist Air Gunner, Petty Officer A. N. Taylor, to render first aid, he died of his injuries aboard the destroyer HMS *Whelp* which rescued the men. Later that day, at 6.45pm, he was buried at sea.

Meanwhile, whilst the strike was in progress, the Japanese made serious efforts to find the fleet and at 9am the enemy reconnaissance aircraft was detected on the radar, but he managed to escape, despite the efforts of the Seafires which were flying Combat Air Patrols. Just over half an hour later the Seafires intercepted a small raid which was approaching the fleet from the north, and one enemy reconnaissance aircraft (Dinah) was shot down. At 10.26am a force of Corsairs and Seafires was directed to attack a group of 12 or more enemy aircraft, and in the ensuing firefight one Corsair was lost. Ten minutes later a group of enemy aircraft passed some 40 miles to seaward of the fleet but, wisely, it was allowed to proceed unmolested. However, just before noon they did locate the fleet and although a force of seven Army twin-engined bombers (Helens) was broken up by the Combat Air Patrols, most of the Japanese planes reached the carrier force and made low-level bombing attacks on the *Illustrious* and *Indefatigable*. Fortunately, they were all destroyed by anti-aircraft fire before they could cause any damage to the ships, but tragically the *Illustrious* was hit by 'friendly' anti-aircraft fire causing 33 casualties, including 12 killed. It was past 7pm that evening before enemy aircraft finally quit the vicinity of the fleet.

One of the strangest tales of the 'Meridian' operations concerns a stowaway in one of 849 Squadron's Avengers. The story is told by Sub-Lt John Brown: 'During carrier take-off in Avenger aircraft the air gunner sat in the gun turret with his protective armour-plated flap in place, so that he was facing to the rear. This was the safest place for him during take-off, but it meant that he couldn't see the inside of the aircraft itself. It seems that as soon as we were ready to go and Chick was in his turret, the stowaway slipped in through the rear door of the aircraft. The first thing I knew of anything amiss was after we were airborne, when Chick went down to his radio station and called to tell me that we apparently had a stowaway on board. We were far too concerned about what faced us on the way to Palembang to bother about a stowaway and I do not think I really took it in until we were safely back on *Victorious* some hours later. Taking a fully loaded Avenger off a carrier is problem enough without the weight of an extra passenger, in addition we had to climb to 13,000 feet to clear the 11,000 feet Barison range of mountains, which is pushing it a bit in a fully laden Avenger. During the trip to and from Palembang, and during our bombing dive, I presume that our guest would have spent most of his time peering out of the small windows of the back cabin. Chick of course was much too busy in his turret keeping look-out and engaging such Japanese fighters as came to have a go at us. For the landing back on *Victorious*, Chick was back in his place in the turret, so he did not actually see our visitor leave. It seems that as soon as we touched down and the aircraft had been released from the wire, the stowaway nipped smartly out of the rear door and disappeared amongst the deck crew. When Chick was questioned officially about the incident, he said the stowaway was a sailor with a red beard, and that he was unable to identify him. In fact, Chick eventually pointed out to me one of our own radio mechanics (clean shaven) and this was later confirmed to me by our engineer CPO Flanagan.'

The two raids on the Japanese oil refineries were the most successful ever mounted by British forces in South-East Asia, with oil production at Soengei Gerong being stopped completely for two months, after which production never increased to more than half capacity. However, losses of aircraft on both sides were heavy with 16 of our aircraft being lost to enemy action, and 25 from other causes. On the Japanese side 30 aircraft had been shot down, and 38 were destroyed on the ground. In both operations 32 British aircrew personnel were lost, and this included nine who were captured by the Japanese and made prisoners of war. All of them were taken by the Japanese Army to Singapore where they were held in the city's Outram Road Prison and on the evening of 18 August 1945, three days after the ceasefire which preceded the Japanese surrender, to the eternal shame of the Japanese they were executed on the beach at Changi on Singapore's north-east shore. Following the reoccupation of Singapore by British forces three Japanese officers, a major, a captain and a lieutenant, were arrested for the murders, for which they admitted their responsibility. However, obviously anticipating the course of justice, all three of them committed suicide, and there was insufficient evidence to implicate any senior officers in the crime. Today there is a memorial stone to the nine men on the floor of the Fleet Air Arm Memorial Church, St Bartholomew's, Yeovilton, and there are memorial plaques in Portsmouth Cathedral and in the Changi Prison Museum, Singapore.

The Royal Navy's official historian has described the 'Meridian' operations as the 'British Pacific Fleet's greatest contribution to final victory over Japan'. This comment leaves a question mark over the importance and value of the British Pacific Fleet's operations off Okinawa and the Japanese home islands.

The *Victorious* arrives at Sydney on Sunday 11 February 1945. With the famous Harbour Bridge in the background and with the Guard and Band paraded while the ship's company man the flight deck, the carrier moves towards her berth at Woolloomooloo.
(C. Minett)

Chapter Nine

Return To The Pacific

After leaving the coast of Sumatra, Force 63 steamed south-east for Australia and refuelled on Tuesday 30 January 1945. Two days later, with the fleet in a position Lat 20°- 42'S/Long 102° - 51'E, and with a much more relaxed atmosphere prevailing on board each ship, a somewhat belated Crossing the Line ceremony was held on the flight deck of the *Victorious*, despite the fact that the ship had crossed and recrossed the equator a number of times already as part of the Eastern Fleet. Ray Barker recalls the event: 'A large canvas pool was constructed on the flight deck. Filled with water it also housed four burly Royal Marines, standing waist deep. Above them and across the pool was a platform on which was balanced a "barber's chair". The "barber", dressed in a long white robe and anti-flash gear, wielded an enormous wooden "razor", while his assistant had a huge container of foaming "lather". Each hapless victim was put in the chair, "shaved", and tipped into the water below where he was seized upon by "Davy Jones'" assistants who immersed their charge three or four times whilst chanting homage to "Neptunus Rex" before dumping him out on the flight deck. Several of the officers were punished for offences against Neptune, even though they were "shell-backs", and Captain Denny was arrested just near the bridge and brought down to be shaved and ducked.'

On Saturday 3 February the fleet skirted the edge of a cyclone which caused some minor structural damage to the smaller ships, but next morning they all arrived at Fremantle and the *Victorious* secured alongside B1 berth at 8am. Although the stopover lasted only just over 24 hours, it was a welcome break for the ship's company and there was an enormous amount of mail waiting to be delivered as well as food parcels from Australian welfare agencies, and tons of fresh grapes. By midday on Monday 5 February the fleet was at sea once again, bound for Sydney, and during the crossing of the Great Australian Bight the aircrew were exercised in the use of American landing signals. Although this led to eight aircraft crashes, including two Avengers

The *Victorious* at anchor off Manus Island in March 1945. *(C. Minett)*

which ditched into the sea, there were no fatalities, although 849 Squadron was very depleted on arrival in Sydney with just 11 aircraft, and only one of them being serviceable. On the morning of Saturday 10 February the battleship *King George V* carried out gunnery exercises, including impressive broadsides with her ten main 14-inch guns. However, in the presence of four powerful fleet carriers it was clear to many observers, who watched the spectacle with some nostalgia, that the battleship was already outdated and that it had been well and truly surpassed as the fleet's capital ship by the aircraft carrier.

At 7.30am on Sunday 11 February the *Victorious* passed through Sydney Heads and shortly afterwards the harbour pilot was embarked. At 9.20am she was secured alongside her berth close to Sydney's Zoological Gardens and most of the ship's company could look forward to an enjoyable 16 days' stay in the port where the hospitality of the citizens was second to none. During the morning of Saturday 17 February Admiral Sir Bruce Fraser made a brief visit to the ship, and the carrier's stay came to an end at 2.35pm on Tuesday 27 February when the *Victorious* slipped her moorings and left Sydney. As she took her departure and sailed out, crowds on the beaches and harbour ferries waved to her enthusiastically.

Once at sea the *Victorious* joined the rest of the fleet, designated Task Force 113 and including the repair ship and ferry HMS *Unicorn*, as they set course for the British Pacific Fleet's forward operating base at Manus Island, the largest of the Admiralty Islands some 3,000 miles from Sydney. The passage took eight days and it was 3pm on Wednesday 7 March when the *Victorious* anchored in the sheltered waters of Seeadler Harbour. Meanwhile, Vice-Admiral Sir Bernard Rawlings, Second in Command of the British Pacific Fleet, was facing huge logistical problems when he discovered that only 40 per cent of the projected Fleet Train had arrived at Manus Island in advance of TF 113. Six days later the *Victorious* put to sea to carry out bombing practice, returning to her anchorage later in the day. Next morning at just before 7.30am, she again weighed anchor to carry out joint exercises with the *Indomitable* and *Formidable*, but at 3.25pm that day there was a serious fire in the centre boiler room. The events are related by PO Ted Martin: 'I took over my watch in the boiler room. It was hot, but soon became much hotter as small fires began. They were quickly extinguished with buckets of sand but, at just after 3pm, there was a massive explosion and the boiler room was engulfed in a mass of flames roaring high into the compartment. I immediately raised the alarm and shut the boiler room down, but there was a mass of flames under the boiler and I grabbed a foam extinguisher and worked my way into the bilges. After what seemed an eternity the flames were at last out, but with the boiler room in a state of partial darkness I was very relieved when damage control parties took over from me.' After half an hour, during which time the centre boiler room was isolated, the *Victorious* was once again steaming in company with the other carriers, and for his fast and courageous action Ted Martin was awarded a Mention in Dispatches. On the following day, at 6.30pm, the *Victorious* anchored at Manus Island and by now the

On 12 April 1945 the *Victorious* experienced her first kamikaze attack when the wing-tip of the enemy aircraft struck the port edge of the flight deck, before cartwheeling over the side where its bomb detonated with a massive explosion.
(C. Minett)

On 9 May 1945 a Japanese kamikaze aircraft crashed into the flight deck where its bomb detonated. This photograph shows the resulting damage to the *Victorious*. (C. Minett)

tropical heat was beginning to take its toll on the ship's company. Ray Barker describes conditions on board: 'The climate was cripplingly hot, with the light winds offering no respite and the temperature below decks was a constant 100°F, with a high level of humidity. The vast flight deck absorbed and retained the heat of the sun and it became almost too hot to walk on with thin-soled shoes or sandals. The heat penetrated into and through the hangar deckhead, and the huge compartment became an enormous sweatbox with the noxious aircraft fumes thrown in. The engine rooms and galleys also generated heat and this seemed to rise upwards from the lower reaches of the ship, to meet that from above. The result was an enervating, energy sapping humidity which made every task an effort and concentration very difficult. The drinking water fountains splashed out warm water, the cold showers were almost hot and sweat seemed to run from every pore as you soaped and rinsed. Drying with a towel was a waste of time and effort. The Manus Island anchorage was a test of endurance and morale for the ship's company.'

On Friday 16 March a tragic accident occurred at 1.10pm when AB J. A. Webber, who was manning the captain's motor boat, fell from a stern boom and was drowned. Two days later there was almost another tragedy when, in the early hours of the morning while one of the cutters was being hoisted aboard, a sling broke and three crew members were thrown into the water and were quickly carried astern by the strong currents. Fortunately, all the ships in the harbour turned on their searchlights and the three men were rescued safely a quarter of a mile away from the *Victorious*. Later that same morning, at 10am, the fleet sailed for Ulithi Atoll in the Caroline Islands, less than 1,000 miles from the island of Iwo Jima. They arrived at the US forward base at just after midday on Tuesday 20 March and three days later were designated Task Force 57. Their commander was Vice-Admiral Sir Bernard Rawlings, who flew his flag in the battleship *King George V* and his Second in Command was Rear Admiral Sir Philip Vian who was also Flag Officer Aircraft Carriers BPF, flying his flag in HMS *Indomitable*.

The British carrier force, though in aircraft complement

barely approximated to a single group in the American Fast Carrier Force, was given the status of a Force in the US Navy's Fifth Fleet. The greatest disadvantage of operating as a single group was that the British Pacific Fleet was prevented from playing its full part in the assault on the Ryukyu Islands, for with replenishment and maintenance to be undertaken it was unable to keep a continuous presence in the operational area. The aircraft carried in the British ships presented another problem as the six operational types required a huge number of different parts, and it was the lack of air stores which came nearer to causing a complete breakdown in operations than any other single factor. When they had left Sydney only ten per cent of their demands for stores had been met. In addition some of the aircraft themselves left much to be desired with their low endurance forcing the whole fleet closer to shore, which in turn meant far too much effort had to be concentrated on defensive measures at the cost of offensive operations. The Seafires in particular were restricted in their range, and they had to be used almost entirely for defensive operations. Although the Avengers, Corsairs and Hellcats were all American-built aircraft they had been so modified for service with the Royal Navy that it was impossible to pool resources with the US Navy.

The first assignment for TF 57 was to support the American landings in the Ryukyu chain of islands which lie south-west of the main Japanese island of Kyushu. The key target for the US landings was Okinawa Shima and the Royal Navy's task was to neutralize airfields on the islands of Miyako and Ishigaki in the chain known collectively as Sakishima Gunto. Although the invasion of Okinawa Shima was set for 1 April 1945, it was important that TF 57's air strikes started prior to that in anticipation of any possible Japanese counter-offensive, and on the morning of Friday 23 March the force left Ulilhi anchorage bound for flying-off positions which were between 80 and 100 miles from the targets. During the first phase of the operation, which had been designated 'Operation Iceberg', the job of TF 57 was to crater the runways on the three primary airfields on both Ishigaki and Miyako, some of which were known to be heavily defended with anti-aircraft batteries. For this operation TF 57 was equipped with 218 aircraft for which the order of battle was as follows: HMS *Victorious*, 37 Corsairs, 14 Avengers and two Walrus (ASR) aircraft; HMS *Indomitable*, 29 Hellcats and 15 Avengers; HMS *Indefatigable*, 40 Seafires, 20 Avengers and nine Fireflies; HMS *Illustrious*, 36 Corsairs, 16 Avengers. It was an extremely powerful force and the first strikes were flown off the *Victorious* at 6.30am on Monday 26 March, when the force was in a position Lat 23° - 50'N/Long 125° - 50'E, about 100 miles south of Miyako. On arrival over the target the strike aircraft made a steep diving attack on Hiraro airfield, Miyako, before swinging round to attack Nobaru airfield on the same island. Although anti-aircraft fire was experienced it was not accurate, although unfortunately, the CO of 1836 Squadron, Lt-Cdr C. R. Tomkinson, was hit and was forced to ditch within sight of land. Despite the fact that he was seen in the water wearing his life jacket, and the fact that both ASR Walrus aircraft carried out a thorough search of the area, no further trace of him was found. It was a sad loss to the squadron and Captain Denny declared, 'He was an experienced Corsair pilot, a first-class commanding officer and a serious loss to *Victorious*.'

At 9.15am the same morning the second strike force of 12 Corsairs and six Avengers was launched to join aircraft from the other carriers to make further attacks on the airfields, with a third strike being launched at just after midday. That night the force withdrew to the south and early next morning they were once again in a position to fly off further strikes, the first of which was launched at 9.12am with the last at 3.10pm. On the second day one pilot, Sub-Lt P. C. J. H. Spreckly was lost and by the end of that day the *Victorious* had launched five strikes against the enemy airfields. However, it was something of a thankless task for those involved and there was some doubt as to whether the results really justified the effort involved, as most of the damage caused to the runways was repaired overnight. The attacks had destroyed about 40 Japanese aircraft, with 28 having been shot down, but TF 57 had lost 19 aircraft. Over the following two days, Wednesday 28 and Thursday 29 March, TF 57 withdrew from the operational area to refuel, but with the American landings on Okinawa due to begin on Sunday 1 April, on the morning of Saturday 31 March the *Victorious* was back in her flying-off position off Sakishima Gunto. There were three strikes that day at three-hourly intervals, with the first being launched at 9.15am. The second strike of four Corsairs and six Avengers took off at just before 12.15pm with orders to attack Ishigaki airfield. They made a good attack, with their bombs hitting both runways, airfield buildings, barracks and a radio station. However, during the raid an Avenger of 849 Squadron was hit by anti-aircraft fire and was seen to crash, with the loss of its crew, Sub-Lt R. C. Sheard, Sub-Lt A. R. Legge and Ldg Airman N. Hawkins.

At 6am on the morning of Sunday 1 April the inexorable American advance towards the Japanese home islands continued when, what was to be the last - and the bloodiest - major amphibious operation of the Pacific War got under way. Two divisions of the US Army and two divisions of the US Marines, a force of 60,000 men, went ashore on Okinawa after an intensive air and naval bombardment. The island was defended by the Japanese 32nd Army - some 130,000 men - and the fighting was fierce, but the American force gradually advanced in the face of formidable defences and suicidal counter-attacks. For the *Victorious* and the other ships of TF 57 it was also a day of continuous action with the force having its first experience of kamikaze attacks, and for the first time the

A second kamikaze aircraft hit the after end of the flight deck, literally seconds after the first aircraft had hit the *Victorious*.
(C. Minett)

The ship's Walrus seaplane returns with rescued aircrew. (C. Minett)

British Pacific Fleet encountered determined resistance. At 6am that morning 'Action Stations' was sounded on board the *Victorious* and just over half an hour later the first Corsairs to be launched set course for their targets on the airfields of Ishigaki. At 6.50am a group of enemy aircraft were detected approaching the fleet from the west, and although they seemed to split up as they approached the fleet, it was clear that there were between 15 and 20 planes rapidly approaching TF 57. Both the strike aircraft and the Combat Air Patrols intercepted the enemy force, but some of the attackers managed to penetrate the protective patrols and at 7.20am, despite the attempts of Sub-Lt R. H. Reynolds to shoot it down, a Zeke suicide bomber dived into the base of *Indefatigable's* island and the massive explosion of the aircraft and its 500lb bomb damaged the flight deck and adjoining compartments, including the sickbay. Eight men were killed and 16 were wounded by the explosion, but although the flight deck was temporarily put out of action, at just after 8am the first aircraft were able to land on. Also damaged in this attack was the destroyer *Ulster* which was disabled and had to be towed back to Leyte. At 8.47am, a Seafire from the *Indefatigable* which had been damaged in the air battles and which was being flown by Sub-Lt N. Quigley, made an emergency landing into the *Victorious'* crash barrier. As he approached the flight deck he was very low and he struck the round down heavily, missed all the wires and crashed into the barrier. The ground crews had great difficulty in getting the pilot out of his cockpit, and as the operation to clear the damaged aircraft from the flight deck was under way its machine-guns fired, but, fortunately, no one was injured. However, the unfortunate pilot died of his injuries and at 4.30pm that afternoon he was buried at sea. Just over an hour later, at 5.38pm, the *Victorious* experienced her first kamikaze attack when a Zeke which had broken through the fighter defences approached from the starboard quarter, and despite a fierce anti-aircraft barrage, his starboard wing-tip struck the port edge of the flight deck abaft S compartment and the whole aircraft cartwheeled over the side and into the sea where his bomb detonated under the water and about 80 feet from the ship's side. The resulting explosion threw huge quantities of water, aviation fuel and debris onto the flight deck, and although two members of the ship's company were injured, the damage to the flight deck itself was only superficial. In addition to casualties sustained during the kamikaze attack, one Corsair failed to return from the strikes on the airfields with the loss of its pilot, Sub-Lt H. J. H. Roberts.

Next day the *Victorious* withdrew from the area in order to refuel, but severe weather hampered the operation and it was the morning of Friday 6 April before the carrier was once again in the operational area and flying off the strike aircraft. However, that morning's flying got off to an uncertain start which is recalled by Ray Barker: 'Standing on the ADP and watching our Avengers take off we had an interesting diversion when, after the first in line had coughed stutteringly into silence, and was pushed away, the second Avenger headed straight for the island. We watched, horror-struck, as the bombed-up plane slewed rapidly off its straight line for take-off and thumped unceremoniously into the base of the island below, and slightly astern, of where we were standing. Fortunately, the bomb did not detonate and the aircrew suffered only minor cuts and bruises, but the strike had to be carried out by the four remaining Avengers. It seemed that the damaged Avenger's starboard brake had jammed, causing it to slew to the right and collide with the island superstructure.'

Later that afternoon, at 5.30pm, there was another enemy raid on TF 57, during which a Seafire was shot down by 'friendly' anti-aircraft fire and the *Illustrious* was damaged by a kamikaze attack when the enemy plane's wing-tip hit the island superstructure before it plunged into the sea. During Saturday 7 April the attacks on the enemy airfields continued, during which one Corsair had to ditch into the sea, with the loss of its pilot, Sub-Lt R. H. Burns. However, his body was recovered by a destroyer and later buried at sea. That evening, as the *Victorious* left the area to refuel, Captain Denny broadcast the very welcome news to the ship's company that the thankless task of bombing the airfields of Sakishima Gunto was over and that for the next phase of 'Operation Iceberg' their target would be the island of Formosa (Taiwan). The first operation was to be against the Shinchiku and Matsuyama airfields in the northern part of Formosa which the Japanese were using as fuelling bases in their attacks on the US ships off Okinawa. The flying-off position was some 50 miles off the island of Formosa, but before the strikes were launched on the morning of Thursday 12 April, the Combat Air Patrols intercepted four enemy Zekes and shot one down. Ten minutes later *Victorious'* strike force of 12 Corsairs and nine Avengers took off for the Formosan airfields and although the raids were successful, one Avenger and a Corsair were lost with their aircrews, Sub-Lts McAlesse and D. McLachlan DSC, Lt F. A. Mavis and Ldg Airman Cloughan. That evening the Japanese made a sortie from Ishigaki, but it was intercepted by fighters before it reached the fleet with eight enemy aircraft being shot down. Next morning the first strike was launched at 6.45am and the report of Lt-Colonel R. C Hay DSO DSC RM, describes the mission: 'We took off to strike Matsuyama Airfield, Formosa, at 6.45am and departed at 7.07am - surely a record. A landfall was made at 7.30am with a visibility of 30 miles. Cloud meant conditions deteriorated over the hills so I ordered the strike to proceed round the coast. This they did, climbing to 8,000 feet. Unfortunately, a solid layer of cloud built up right over the airfield and prevented a detailed examination of the target. The cloud layer was 1,000 feet thick with a base of 3,000 feet. Informed the strike leader, D. R. Foster, and he orbited overhead looking for a gap - he eventually decided to bomb after diving through cloud. From down below I observed many bombs striking the airfield until it was covered in brown smoke and dust. There was no flak before the bombing, but the moment the bombers appeared below 3,000 feet every gun went into action. They were obviously waiting. On withdrawal one Avenger bombed a factory. I caught a passenger train which was skulking in a tunnel but rather carelessly had left the engine sticking out. We then shot up a few junks and went on to Giran where approximately 12 aircraft of various types were spotted. One twin-engined plane was strafed but failed to burn.' Given the weather conditions Lt-Colonel Hay considered the strike to be a success. That same day four enemy aircraft got through the fighter defences to make an ineffectual attack on the *Indomitable*, and on Saturday 14 April the *Formidable* joined the fleet off Formosa. The *Victorious* made her final strikes of the phase on Friday 20 April, and that evening TF 57 left the area for a well-earned rest and maintenace period at Leyte, arriving at San Pedro Bay in line ahead at just after midday on Monday 23 April.

The *Victorious* had been at sea continuously for some 31 days since leaving the anchorage at Ulilhi on Friday 23 March and the strain on everyone, particularly the aircrews, was beginning to tell. Although leave was granted to the ship's company, to reach the shore meant a long and difficult journey by tender, and as Les Vancura noted in his diary: 'We have beer on board for the first time and each man has one and a half pints at midday and one and a half pints at 5pm.' Entertainment could be found in a boxing tournament and a comedy Sports Day which were held on the flight deck.

It was Tuesday 1 May 1945 when the fleet left San Pedro Bay to return to the operational area, with the *Formidable* replacing the *Illustrious*, and once again they were under orders to continue with the neutralization of the Sakishimi Gunto airfields, with a continual cycle of two days of strikes followed by two days of replenishment. On Thursday 3 May, as the force neared the operational area, the ships fuelled and early next morning the first strikes were flown off. At 10.08am the battleships and cruisers detached from the fleet to carry out a bombardment of the Japanese airfields and anti-aircraft positions on Miyako. This was partly for the benefit of the young and untried members of the ships' companies, but with the carriers remaining some 30 miles to the south the Japanese seized the opportunity to bring them under air attack by some 16 to 20 aircraft working with decoys. There were no enemy aircraft showing on the radar screens however, when at 11.33am a Zeke dived down on the *Formidable* and was engaged by the vessel's close-range weapons. Despite being hit a number of times, the kamikaze smashed into the flight deck near the island starting a large fire in the parked aircraft, causing a large number of casualties and damaging the flight deck to a depth of two feet. Two minutes later another kamikaze hit the *Indomitable*, but it bounced over the side without causing any serious damage. By mid-afternoon the heavy units had returned to the fleet and several enemy aircraft had been shot down. At 5pm the *Formidable* was operational again following the major damage which she had sustained and shortly before the fleet withdrew for the night another attack developed, but this was successfully broken up by the fighters. On Saturday 5 May further strikes were flown, and then there were two days of replenishment before the *Victorious* returned to undertake more strikes on 8 May, while the two battleships *King George V* and *Howe* carried out a bombardment with their main armament.

On Wednesday 9 May the first strikes were flown off at 10.42am and five hours later the fourth strikes were launched to attack the airfields of Sukishimi Gunto. However, at 4.47pm the radar operators on board the battleship *King George V* reported enemy raiders at 28 miles and soon after this they were picked up on the *Victorious*' radar. Nine minutes later one enemy aircraft was reported nine miles away and Captain Denny ordered an emergency turn of 60 degrees to starboard. At almost the same time the raider was sighted on the starboard quarter, at 3,000 feet and about three miles away. All the ship's anti-aircraft guns and close-range weapons opened up as the aircraft closed. By now it was clearly heading for the *Victorious* and a 25 degree turn to starboard was made, but despite direct hits to the aircraft and smoke being seen coming from it, the damage was not sufficient to prevent the suicide bomber from completing his attack. The kamikaze came in a shallow dive past the bridge, crossed the ship and crashed into the flight deck on the accelerator abaft 'B Group' where its bomb detonated with a massive explosion.

Hardly had Captain Denny steadied the ship when a second kamikaze was sighted on the starboard quarter and once again all the anti-aircraft guns opened fire. Although the aircraft was hit and flamed at about 500 yards, it continued its dive in from starboard and skidded on the flight deck before bouncing into the sea about 200 yards on the port beam, having cut right through the Corsairs which were parked aft on the flight deck. Then, four minutes later at 5.30pm a third suicider was sighted on the starboard beam at 3,000 feet and about 3,000 yards away. Once again this aircraft was brought under heavy fire from every available weapon as it flew up the starboard side before it turned to dive on the battleship *Howe*. Fortunately, this one was flamed about 800 yards from the *Howe* and it flew straight over her to crash harmlessly into the sea. Very shortly after this, the fourth and last suicider was sighted on *Victorious*' starboard quarter and once again the anti-aircraft guns on that side opened fire. However, as the aircraft began diving towards the *Formidable*, fire had

The *Victorious* refuels the destroyer HMS *Wager*. *(C. Minett)*

Tuesday 25 September 1945 and the *Victorious* leaves her berth at Woolloomooloo, Sydney, to return home to the UK. *(C. Minett)*

to be checked for fear of hitting the carrier and the plane dived into her after deck park, putting 18 aircraft out of action but causing very little other damage.

On board the *Victorious* the first suicider had crashed near to the forward lift where its bomb had detonated, but the resulting fire was quickly brought under control. However, the explosion had holed the flight deck and damaged an electric motor. The second suicider had damaged four Corsairs, but it had caused very little structural damage to the ship. Sadly, three members of the ship's company, Steward J. Landers, and Able Seamen A. A. Bliss and J. E. K. Cann, were killed in the attacks. There were also 20 wounded, four of them seriously, and one Corsair failed to return from the final strike which had flown off at 3.42pm, with the loss of its pilot, Lt D. Cameron of 1834 Squadron.

Over the next two days TF 57 withdrew for replenishment and when operations resumed on Saturday 12 May repairs to the flight deck and to electrical fittings had been completed. That day a Corsair returning from a fighter sweep was forced to ditch in the sea and the pilot, Sub-Lt B. J. Smith, was very fortunate to be rescued having reported his position incorrectly, being picked up after he managed to signal to passing Fireflies with a mirror. The next incident to affect the *Victorious* took place on 16 May when a Corsair ditched in the sea, with the pilot being rescued by the American submarine USS *Bluefish*. Next morning, at 5.38am, the first strike of four Corsairs and four Avengers was launched, and they returned to the carrier at 7.40am. One Corsair which had been damaged by flak arrived back with its flaps jammed and the pilot, Sub-Lt K. W. Hardiman, made a very fast landing which parted No 6 and No 8 wires. The aircraft then careered along the deck straight into the rear Corsair in the forward park, before it plunged over the side in a ball of fire. The unfortunate pilot, together with Lt W. A. L. Banning and Stoker Petty Officer E. H. Groves, who was in charge of the crash barriers, were all killed in the incident. The resulting fires were soon extinguished, but it was not until 11.30am that the flight deck was once again ready for use. At 11.48am two Corsairs were landed on, and the second unfortunately landed so as to hook at the vacant No 6 wire. The hook then bounced over No 7, tried for the vacant No 8 and then bounced over Nos 9 and 10. The aircraft then careered into a weakened after crash barrier which stopped the aircraft, but completely wrecked the barrier mechanism. Thus the after crash barrier was irrecoverably out of order. That afternoon a jury barrier which had been rigged was wrecked when another Corsair missed the wires. In the event 20 aircraft from the *Victorious* had to land on other carriers until repairs were effected.

Next day it was the *Formidable's* turn to suffer a serious accident. Whilst fuelling and exchanging aircraft a Corsair in the hangar accidentally fired its guns into an Avenger which exploded. The resulting fire destroyed or damaged 30 other aircraft, but by the evening the ship was operational once more. On Saturday 19 May the *Victorious* carried out an ammunition RAS with the supply ship *Robert Maersk*, during which she embarked 500lb bombs at an average rate of 81 per hour which, although not as fast a rate as the US Navy could achieve, was extremely efficient as this kind of under way replenishment had not been tried before the ship joined the Pacific War. Over the following four days two more pilots, Lt-Cdr J. B. Edmunson DFC and Sub-Lt A. C. Ralston of 1836 and 1834 Squadrons, were lost. The former on Sunday 20 May when his aircraft crashed into the sea whilst returning from a sortie and the latter at 7.14am on Wednesday 23 May, whilst landing on. His approach to the carrier was far too low, and his aircraft hit the round down and fell into the sea. Sunday 20 May had been very foggy and the destroyer *Quilliam* had collided with the *Indomitable*, badly damaging her bows. In the event she was taken in tow by another destroyer, and with the cruiser *Black Prince* as escort she was towed to the fuelling area at 5 knots. The final strikes in 'Operation Iceberg' were made on Friday 25 May, and that evening, whilst approaching to land on, a Corsair ditched in the sea. Fortunately, the pilot, Sub-Lt Barlow, was rescued by the destroyer *Quickmatch*.

Next day TF 57 set course for Manus, en route to Sydney where the *Victorious* was to undergo a refit and prepare for operations which would involve the US landings on the Japanese mainland. Both Admirals Spruance and Halsey had paid tribute to the excellent work carried out by TF 57 and during 'Operation Iceberg' the British carriers had flown 4,961 sorties, and had dropped 927 tons of bombs. In the five months which had passed since her dry docking the *Victorious* had steamed some 42,500 miles and the defects resulting from the prolonged periods at sea were increasing, in addition to which all the ship's company and squadron personnel were in need of a break.

The *Victorious* arrived at Manus on the morning of Thursday 31 May, and next day at 1.30pm she weighed anchor and set course for Sydney. During the voyage south she encountered some very severe weather when she skirted the edge of a typhoon, but during the morning of Tuesday 5 June she reached Sydney Harbour and secured to a buoy before shifting to No 7 berth Woolloomooloo that same afternoon where she secured alongside at 5pm. During the 21-day stay the damage sustained in the kamikaze attacks was repaired, a draft of 142 ratings joined the ship and the ship's company enjoyed the hospitality of the city. The *Indomitable* went into dry dock, whilst the *Indefatigable* and *Formidable* were also berthed at Woolloomooloo, but much to the envy of the other ships' companies the *Illustrious* had left for home. Les Vancura recalls in his diary that during the time spent in Sydney he attended a Gracie Fields concert, visited the Blue Mountains and the resort of

Katoomba. On Tuesday 19 June the Governor-General of Australia, HRH The Duke of Gloucester, paid a short visit to the ship and on Sunday 24 May a contingent of War Correspondents embarked as, once again, the *Victorious* prepared for sea. It was at just after midday on Monday 25 June that she slipped her moorings and left Sydney to steam round to Jervis Bay, where her squadrons were flown on. For three days she carried out flying exercises in the area, before leaving for Manus where she arrived, with the *Formidable* and *Implacable*, on Wednesday 4 July. During these final operations against Japan the British Pacific Fleet was designated Task Force 37, and the four fleet carriers were as follows:-

Formidable: (flagship Vice-Admiral Sir Philip Vian) - six Hellcats; 36 Corsairs; 12 Avengers.

Victorious: 37 Corsairs; 14 Avengers; two Walrus ASR.

Implacable: 48 Seafires; 12 Fireflies; 18 Avengers.

Indefatigable: 40 Seafires; 12 Fireflies; 18 Avengers.

In command of the fleet was Vice-Admiral Sir Bernard Rawlings, who was flying his flag in the battleship *King George V*. The fleet was under the overall command of Admiral Halsey who would work in close co-operation with the US Navy's TF 38, which was split into three task groups operating 16 aircraft carriers of which ten were fleet carriers. In practice the British Pacific Fleet would form a fourth Task Group of TF 38.

The fleet left Manus on Friday 6 July and ten days later, in a position Lat 39° - 10'N/Long 148° - 30'E, it rendezvoused with TF 38, approximately 100 miles due east of the main island of Honshu. Once again the main duty of the Fleet Air Arm was to attack enemy airfields, but also any other targets which presented themselves, and operations began at 3.47am on Tuesday 17 July when the *Victorious* launched 16 Corsairs for airfield strikes. During flying operations the fleet closed to within 70 miles of the Japanese coast and on occasions the battleships moved in very close to bombard various targets. By now deficiencies in the British Fleet were beginning to make themselves felt and the greatest difficulty was experienced in keeping our ships supplied with fuel and essential spares. The fuelling problems arose because of the inadequacy of our tankers and their low rate of pumping. The US Navy's aircraft carriers were faster and bigger than our fleet carriers and their aircraft were newer. It was clear that if the war against Japan were to last for a prolonged period then the British Pacific Fleet, despite the heroic efforts of all involved, would be unable to stay the course with the Americans. However, with the typhoon season in the area the continual bad weather caused flying to be cancelled time after time, which gave TF 37 some breathing space.

Admiral Halsey denied the British Pacific Fleet any participation in the destruction of the Japanese Fleet in order to forestall any post-war claim by Britain that she had taken even a small part in this operation, and the British targets continued to be enemy airfields. On Monday 6 August 1945, when the US Air Force dropped the atomic bomb on the city of Hiroshima, the British Fleet had been withdrawn for refuelling, but three days later when the second atom bomb was dropped on Nagasaki, the fleet was carrying out strike operations. The *Victorious* started flying off strike aircraft at 4.10am and sadly that day she lost one Avenger and its crew. Next day, when the Japanese Government sued for peace, the *Victorious* was once again flying off strikes against shipping at Okkaichi and Onagaw Wan, and airfields in the area. For the British Pacific Fleet these were the closing stages of operations and 227 of the Fleet Air Arm's aircraft attacked the targets. Sadly six aircraft with six aircrew were lost in combat that day, but amongst the targets destroyed were an auxiliary minesweeper and a patrol vessel. Next day the fleet withdrew for refuelling and because of severe weather all strikes which had been planned for 12 August were cancelled. Once Japan's surrender became imminent the British Pacific Fleet split up, with the *Indefatigable* and other units joining the US Navy to take part in the enforcement of the capitulation while the other carriers, including the *Victorious*, returned to Manus. Finally, at 9.10am on Wednesday 15 August 1945, with the *Victorious* in a position Lat 20° - 45'N/Long 156° - 30'E, came the news which everyone was longing for - the war was over. Captain Denny announced the news by tannoy. It was a great day for all those on board the ship with a Make and Mend being granted to all except those on watch, and at 6pm 'Splice the Main Brace' was piped. However, it was not until 9.29pm that evening that the official signal was received from the Admiralty ordering all offensive operations against the Japanese to 'cease forthwith'. Next day, as the *Victorious* steamed south, Admiral Vian transferred from the *Formidable* to thank the ship's company for their magnificent efforts. Referring to the *Victorious* as the 'Trojan' he was full of praise for the ship's company who had missed only one minor operation against the enemy since she arrived in the Eastern Fleet.

The *Victorious* arrived at Manus on Saturday 18 August and 24 hours later she left for Sydney and, it was thought, a refit. Five hours after sailing, at 6.40pm on Sunday 19 August, as the *Victorious* was steaming south, her rudder jammed when it was hard over to starboard. Fortunately, it was soon working once again, but the problem recurred twice on 22 August. On Thursday 23 August the squadrons were flown off to an airfield near Maryborough, 100 miles north of Brisbane, and at 1.35pm the *Victorious* anchored in Moreton Bay where the ground crews were ferried ashore in a landing craft. Less than an hour later the carrier was under way once again, and during the afternoon of Friday 24 August she passed through Sydney Heads. The welcome into the harbour is described by Les Vancura: 'Almost every vessel, merchantmen, small warships, ferries,

On the morning of Saturday 27 October 1945 the *Victorious* steamed past the Round Tower into Portsmouth Harbour, after spending 16 months with the Eastern and Pacific Fleets. *(C. Minett)*

yachts and motor boats sounded their sirens or hooters, and some fired flares. Planes flew overhead and dipped their wings in salute.' By 4.30pm the carrier was secured alongside 5b berth of Clarke Island. The war was truly over and few members of the ship's company felt any envy for those on board the *Indefatigable* in Tokyo Bay, for the citizens of Sydney were celebrating in style and the *Victorious* was to be part of the celebrations. There were now other priorities in the minds of most members of the ship's company and these were getting home and demobilization.

Three days after arriving in Sydney, at 9am on Monday 27 August 1945, Captain John Campbell Annesley DSO RN took over the command of the *Victorious* from Rear-Admiral Denny who had been promoted and appointed Chief of Naval Personnel at the Admiralty. He had already said his farewells to the ship's company on the previous day when he thanked them for all the hard work carried out during his time in command, and let them know that he had told Admiral Vian that they were the best ship's company in the British Pacific Fleet. Ray Barker recalls that at the end of his speech, 'Two thousand voices were raised in a cheer of such volume that it rang and echoed round the dockyard.' He finally took his departure at 4pm on 27 August, once again to the cheers of the ship's company.

The *Victorious*' new commanding officer, Captain Annesley, was a Portsmouth man and he had served in the Great War of 1914-18, being wounded in April 1918 during the raid on Zeebrugge. For his part in the operation he was awarded the DSO and the Croix de Guerre. During the Second World War he had been Mentioned in Dispatches and had received the Royal Norwegian Order of St Olav.

The *Victorious* was to be his last command before retirement.

On the same day that he took command Captain Annesley was able to announce to the ship's company that the proposed seven-week docking period in Sydney had been cancelled, and that the ship would soon be leaving for home. However, with each watch having been granted four days' leave, it was clear that she would not be leaving immediately. On Friday 31 August there was a Victory parade through the city centre of Sydney, with contingents from the *Victorious*, *Formidable* and *Implacable* amongst those representing the British Pacific Fleet and with the salute being taken by the Lord Mayor of the city. Ray Barker recalls that as they marched proudly down the main streets, 'Young women broke from the crowd and embraced and kissed us. Shredded paper and streamers thrown from the higher windows covered us in giant confetti.' Meanwhile, on board the *Victorious* work was under way to build dormitory accommodation in the hangars, which were separated by partitions, with a canteen provided at one end. A new rudder had been shipped out to Sydney in order that it could be fitted during the period in dry dock, but with the refit having been cancelled there was the problem of what to do with the massive object. In the event it was decided that the *Victorious* herself would ship it home, and it was lifted onto the flight deck and welded down, in company with numerous crates of aircraft parts and even complete planes, all lashed down for the voyage. On 18 September, 20 officers and 342 ratings from the *Formidable*'s ship's company joined the *Victorious* for the voyage home, as well as a contingent of Far Eastern prisoners of war who had been brought from Japan in the

Indefatigable. It seemed that the ship was being prepared for the next phase of her career - as a troopship. Finally, at 10.30am on Tuesday 25 September 1945, came the moment for which everyone had been waiting, when the *Victorious* slipped her mooring ropes and set course for home. The occasion is described by Ray Barker: 'Hundreds had assembled to wish us farewell. On the flight deck the Royal Marines Band played "Now is the Hour" and on the jetty there were crowds of cheering, waving, crying friends and newly found sweethearts with tears streaming, as the rust-stained *Victorious* moved slowly down the harbour.'

On Monday 1 October she called at Fremantle to refuel and just over 24 hours later she sailed for Colombo, running into severe squalls and heavy seas. During the passage to Ceylon there was further trouble with the rudder and steering gear, but on the afternoon of Tuesday 9 October she anchored safely in Colombo Harbour for a stay of less than 48 hours. On leaving Ceylon the *Victorious* steamed non-stop to Port Suez where she arrived on the morning of Thursday 18 October. All that day was spent on the northbound passage of the Suez Canal, and at 9.30pm that evening the carrier moored in Port Said Harbour. Next day, whilst the ship was fuelling, leave was granted and it was to be the last foreign run ashore of the long commission. At just after 9.30pm on Saturday 20 October the *Victorious* left Port Said for Portsmouth and after a 12-hour stop at anchor in Gibraltar Bay, and a very rough crossing of the Bay of Biscay, the *Victorious* arrived off St Catherine's Point on the Isle of Wight at 5.30am on Saturday 27 October 1945.

Although she had been expected to enter harbour at 9am that morning she was kept waiting off the Isle of Wight and finally, at just after 9.30am and preceded by an Admiralty paddle tug, the *Victorious* steamed majestically through the narrow buoyed channel skirting Southsea seafront to the cheers of onlookers assembled at Clarence and Victoria Piers. The carrier was flying her paying-off pennant while the ship's company manned the flight deck and the Royal Marines Band played, 'I Do Like To Be Beside The Seaside'. As she approached the harbour mouth, salutes were exchanged with Fort Blockhouse and at 10.25am, to the delight of a large group of relatives on the dockside, the *Victorious* secured alongside South Railway Jetty in Portsmouth Dockyard.

In his report to the C-in-C Plymouth, Captain Annesley describes the long voyage home thus: '*Victorious* left Sydney on 25 September 1945, being still in full commission, and opportunity was taken to increase the number of persons on board to 2,600 by the addition of passengers due for release. The voyage cannot be described as a trooping trip, since the old ship's company had shaken down in the ship and many of them had holes and corners in which to sleep. Consequently the overcrowding was not apparent. Morale was very high and the homeward trip was undoubtedly enjoyed by everyone on board. A very comprehensive programme of sports and entertainments was arranged and there was hardly a dull moment. Also, the Royal Marines Band was still in the ship and this undoubtedly contributed very materially to the good time that was had by all.'

The commission had lasted for over four and a half years and the *Victorious* had steamed 268,000 miles during that time, which was the equivalent to 50 years of peacetime steaming. During that time most of the ship's company had stayed with the ship while she operated in every ocean except the Antarctic. She had sailed round the world and had taken a major role in every operation, bar one, which had involved the Eastern and British Pacific Fleets. Now a very different role lay ahead for her.

Friends and relatives pack South Railway Jetty in Portsmouth Dockyard to greet the *Victorious* upon her return from the Pacific. *(C. Minett)*

Chapter Ten

Troopin' And Training

Following her return to Portsmouth on 27 October, most of the ship's company were able to get home for at least a long weekend, but the paying-off process itself was carried out in two phases. The first stage took place at Portsmouth on Sunday 28 October 1945, and the whole procedure was concluded at Devonport some weeks later. In the immediate future the ship was scheduled to undergo her much delayed docking period, this time in Devonport Dockyard and so, at 3.30pm on Monday 30 October, she left Portsmouth and steamed out to Spithead where she anchored for the night. Next morning, at just before 6am, she weighed anchor and set course for Plymouth. During the forenoon, as the *Victorious* steamed along the south coast, the Navigating Officer gave the ship's company a running commentary over the tannoy on the places of interest as they passed by, and at just before 3pm on the last day of October the ship tied up to a buoy in Plymouth Sound where she remained for the night. Next day, at 2.30pm, the *Victorious* slipped her moorings and without ceremony she steamed up-harbour where, an hour and a half later, she was secured alongside the east wall of No 5 basin in Devonport Dockyard, opposite the partially completed light fleet carrier, HMS *Terrible*.* Les Vancura recalls going ashore that evening for a slap-up supper of steak and chips, followed by jam tart and custard - clearly, it was good to be home.

At noon on Friday 2 November the *Victorious* was manoeuvred into No 10 dry dock where destoring and deammunitioning began, and work started to replace the ship's rudder. Later that month, on Wednesday 28 November 1945, phase two of the paying-off process was completed and most members of the ship's company had now left the ship. The new drafts, who were mainly pre-war regulars, were going to find life on board very different from that which had been experienced under wartime conditions and, indeed, the duties of the *Victorious* were to be more suited to those of a troop transport.

When the Second World War ended, Britain, although victorious, was virtually bankrupt. For six years the country had poured its entire industrial capacity and wealth into the war effort and there would be as many problems recovering from victory as Germany would face recovering from defeat. The British Mercantile Marine was now some 30 per cent smaller than it had been at the beginning of the war, yet to survive Britain had to export far more than she had done in pre-war days. There were also vast armies which had been raised both in Britain and in the Commonwealth countries and these troops had to be repatriated, and with Britain's troopships full to capacity, for this enormous task it was decided to utilize the fleet aircraft carriers, which had formed the British Pacific Fleet, to carry both civilian passengers and service personnel between Britain, South-East Asia and Australia.

The *Victorious* left the dry dock during the morning of Thursday 13 December 1945, and later that day the main draft of passengers embarked for their voyage east in the carrier. For her passenger carrying role the ship's company had been reduced to 595, made up of 53 officers and 542 ratings, and she embarked 2,024 passengers whose numbers were made up as follows: 1,829 service passenger ratings, 23 dockyard civilians, three civilians messed as officers, 69 RAAF officers, six Chinese naval officers and 94 naval passenger officers. In addition the ship was also transporting 54 Seafires, six Fireflies and two Walrus aircraft, all of which were lashed down on the flight deck as the hangar dormitories were fully occupied by passengers. With sailing having been delayed for 24 hours because of fog, it was 1.28pm on Saturday 15 December 1945 when the ship left Devonport and after passing the breakwater of Plymouth Sound an hour later she set course for Gibraltar at a speed of 20 knots. However, next morning when she was off Cape Finistère southerly gales and heavy seas were encountered, and in order to ease the motion of the ship and to reduce the risk of damage to the aircraft which were parked on the flight deck, speed was reduced to ten knots. Next day, with the weather having improved, speed was increased and on Tuesday 18 December the carrier anchored in Gibraltar Bay to refuel. Four days later she arrived in Port Said, but her southerly transit was delayed because of thick fog in the lower reaches of the canal and leave was granted to the passengers and ship's company. In the event the *Victorious* passed through the Suez Canal during the daylight hours of Sunday 23 December, and both Christmas Eve and Christmas Day were spent steaming south through the Red Sea, to arrive in Aden on Wednesday 26 December. Ray Barker, who was one of the few wartime members of the ship's company still with the *Victorious*, describes life on board: 'The voyage became a cruise, the regime almost pleasurable. No "Action Stations" and very little cypher traffic, in fact working days were

---

*HMS *Terrible*, a light fleet carrier of the Majestic class, which had been built at Devonport Dockyard. She eventually commissioned as HMAS *Sydney*.

On the afternoon of 1 November 1945 the *Victorious* steamed to Devonport for the second phase of her paying off and for a refit. Here she is shown arriving at Devonport Dockyard.
(Maritime Photo Library)

strictly 8.30am to 5pm. Time to sunbathe once we had entered the Red Sea with warm weather from then on.'

After leaving Aden in the early hours of Thursday 27 December course was set through the Arabian Sea for Colombo and during the passage steering trials were carried out with the newly fitted rudder. No problems were experienced and at 12.30pm on the last day of 1945 the *Victorious* arrived in Colombo. During the 48-hour stopover the ship was fuelled to capacity and a large draft of 750 ratings, 27 officers and 22 civilians disembarked for service with the East Indies Fleet. After the New Year celebrations at Colombo the *Victorious* left the harbour during the afternoon of Wednesday 2 January 1946 to make an overnight passage round to Trincomalee in waters which were so familiar to the ship just 12 months earlier, but already it seemed a lifetime away. At Trincomalee 51 Seafires, the Fireflies and Walrus aircraft were disembarked, an evolution which was carried out in just nine hours. The aircraft were originally to have gone to HMS *Indefatigable* at Singapore, but in view of the fact that she too was repatriating service personnel they were diverted to RNAS Trincomalee. Having disembarked the aircraft in Ceylon it

HMS *Victorious* enters Colombo Harbour on 9 February 1946, on her return leg of the first post-war trooping voyage.
(C. Heath)

The *Victorious* anchored in Plymouth Sound on 7 August 1946, following her arrival with the Australian 'war brides'.
(Maritime Photo Library)

was not necessary for the *Victorious* to call at Singapore, and so at just after 6pm on Thursday 3 January she sailed for Fremantle. Four days out of Trincomalee the carrier 'Crossed the Line' with due ceremony on 7 January and three days later she anchored in Gage Roads at Fremantle, just long enough to embark fresh provisions before leaving for Sydney. It was on the morning of Tuesday 15 January that the *Victorious* anchored in Sydney Harbour, before securing alongside at Woolloomooloo when a berth became available later in the forenoon. Once alongside, the main draft of passengers disembarked and preparations were made to embark the next complement.

With a number of Wrens having been booked for the passage home, the Australian Dockyard workers constructed cabins for them on the main hangar deck which took nine days to complete, and it was during the afternoon of Friday 25 January that the *Victorious* slipped her berth at Woolloomooloo with 1,512 passengers, which included 31 officers, 1,344 ratings, 117 Wrens (six officers and 111 ratings), and 20 Nursing Sisters and VADs. Included among the naval ratings taking passage home were three offenders who were in custody after being sentenced to periods of imprisonment which they were to serve in the UK. Also embarked as deck cargo were 20 Seafires, three of which the carrier had brought out from the UK, and 183 tons of sheet iron and machine tools which were bound for Singapore. The most important cargo was 75 tons of 'Food for Britain' stores, with a further 20 tons being loaded in Fremantle on 30 January. The *Victorious* arrived in Singapore's Keppel Harbour, on the east side of the city, on the afternoon of Monday 4 February, where 14 officers and 23 ratings were disembarked and 35 ratings embarked for the passage home. Next day the *Victorious* steamed round the coast of Singapore Island to berth alongside at the Naval Base where the engineering stores were unloaded and two Japanese field guns were taken on for transportation to the UK. It was at 7am on Wednesday 6 February 1946 that the carrier left Singapore, and sailing home via Colombo, Aden, Port Said and Gibraltar, she arrived alongside No 5 berth in Devonport Dockyard on Wednesday 27 February where all her passengers disembarked and the precious food cargo was discharged. In his report to the C-in-C Plymouth Captain Annesley gave a summary of the round voyage: '*Victorious* left Plymouth on 15 December on her first real trooping trip with a total of 2,619 passengers on board. Despite the fact that many were young and were leaving the UK for the first time, and that within a few days of Christmas, it can be said that the voyage to Sydney, despite almost continuous bad weather, was a success.' Captain Annesley stressed the need for a band in the ship so that some sort of entertainment could be provided during non-working hours, although on the homeward passage it was clear that there were some musicians on board for he wrote: 'Dances were held on the flight deck on several evenings at sea in the Indian Ocean, which inevitably had an excellent effect on morale. The highlight was undoubtedly the fancy dress carnival and dance just after leaving Aden for Suez - this had strong support from everybody and the number of entries and the standard of disguise were both amazingly high. The Wrens and the dance band contributed a major share to the success of the evening.'

On her arrival in Devonport the ship's company were able to take leave and the ship herself underwent some much needed maintenance. Despite this, however, during the evening of Sunday 10 March 1946 she suffered a fire in a boiler room, but it was brought under control in less than an hour and was described by the Admiralty as 'quite a minor affair'. Certainly it did not delay the *Victorious*' next voyage, for at 4pm on Friday 15 March she left Devonport once again bound for Sydney. This time, not only was she carrying service passengers, but she also had 34 commercial civilian passengers who were being sponsored by the Australian Government. They were some of the first Assisted Passage Migrants or, as they became known, 'Ten Pound Poms', because their only contribution towards the cost of the passage was ten pounds, with the balance being paid by the Australian Government. In those class-conscious days these passengers obviously caused some tensions on board and Captain Annesley's comments make quaint reading today: 'The passengers sponsored by Australia House were a very mixed crowd, and included a labourer and a tramway motor man. These were not men who had suffered a reverse of fortune, for they had followed these callings all their lives, yet they had to be accorded the status and occupied accommodation equivalent to officers of the rank of lieutenant commander. This was in accordance with insructions from Australia House.' Nevertheless, Captain Annesley was able to conclude his report thus: 'With such a motley assembly all living together in the wardroom it was some time before everyone settled down. Once, however, this initial stage was over, nearly everyone had a very enjoyable trip.' In addition to the migrants there were 42 naval officers, three Greek naval officers, 35 RAAF officers, 16 Army officers, 17 clergymen, two Merchant Navy officers and 75 ratings, making 225 passengers in all. Also embarked were 36 aircraft engines which were destined to be dumped in the sea.

Once again the *Victorious* sailed by way of Gibraltar (with the aircraft engines being ditched off Europa Point), Suez, Aden and Colombo where she moored in the harbour at 12.07pm on Monday 1 April. At Colombo some of the passengers left the ship to take up their posts and 20 aircraft along with 96 aircraft engines were loaded for dumping at sea. After embarking 18 more officers, 48 ratings and two and a half tons of fresh vegetables, the carrier left Colombo during the afternoon of Wednesday 3 April. One hour later she stopped for a short while for the redundant aircraft and engines to be ditched over the side, and seven days later she arrived in Fremantle. On arrival at Sydney on Monday 15 April all the passengers were disembarked and a conference was held to determine the date of sailing for the UK, and to plan the embarkation of homeward bound passengers. However, as the ship was to be alongside for at least seven days, the opportunity was taken to give 48 hours' leave to each watch over the Easter period. Planning the embarkation was not an easy task for, with the war over but with many thousands of conscripts still on overseas service, there were considerable numbers of disaffected men who had refused to sail in the carriers. Some of them disliked the accommodation on board, despite the fact that it was not a great deal different from that found on the troopships, and some simply did not wish to leave Australia. In fact, during the time that the *Victorious* was in Sydney, the authorities were worried that there might be a mass breakout from the transit camps in New South Wales. Fortunately, the *Victorious* was able to embark 1,580 passengers without incident, including 1,349 naval ratings, 14 naval officers, 94 Wrens and 78 Royal Marines, as well as two trucks and three motor cycles, and 43 VADs. As deck cargo she carried six Corsairs, six Fireflies, six Seafires and one Walrus aircraft for delivery to Ceylon when she sailed for Fremantle on the afternoon of Tuesday 23 April. During the crossing of the Great Australian Bight she steamed into severe gales and heavy seas, and flooding in the bow section of the ship indicated that she had suffered underwater damage. On arrival at Fremantle on Sunday 28 April divers were sent down to inspect the hull and they found that a large section of the ship's stem and bow plating was missing, leaving a hole measuring 12 feet by 8 feet. It was thought that sections of the hull which had been weakened in the Pacific War had been forced off by the heavy seas and it was decided the ship would have to be dry docked in Sydney. As soon as it was realized that the ship would be considerably delayed, the C-in-C British Pacific Fleet arranged to alter the schedule of HMS *Indomitable*, which was on her way to Australia from the UK. The *Indomitable* arrived in Fremantle in early May and it took two days to transfer all the passengers and cargo from the *Victorious*. The opportunity was also taken to exchange some of the ship's company, who were overdue for release, with men from the *Indomitable*. During the enforced stay in Fremantle every opportunity was taken to encourage local people to visit the ship, and on four open days over 25,000 people took advantage of the offer to look round. There was also a massive party for 1,000 local children who were entertained to tea on board. Both watches were granted a week's leave, and as well as carrying out temporary repairs to the hull, which entailed pouring hundreds of tons of concrete into the damaged section, alterations were made to the hangar accommodation in order that at least 500 wives of naval personnel could embark for the passage to the UK. Finally, at 2pm on Thursday 30 May, with her temporary work having been completed, the *Victorious* left Fremantle and by the morning of Friday 7 June she was high and dry in the Captain Cook dry dock in Sydney Harbour.

During the ship's stay at Sydney platoons from the ship's company took part in the Victory Day Parade through the city, whilst many others took weekend leave breaks. Whilst the repair work was being carried out on the hull, further alterations were made to the hangar accommodation so that a further 200 'Australian War Brides' could take passage to the UK. With the British Pacific Fleet having been stationed in Sydney for well over a year it was inevitable that many men would marry local girls and with their husbands now returning home the new wives wished to be reunited with them. The *Victorious* left the dry dock on Friday 28 June and four days later the embarkation of passengers began. In the ship's log the event was recorded thus: 'Commenced embarkation of wives of naval personnel for passage to UK.' On the following afternoon, having embarked 971 passengers, which included 619 'war brides', 318 naval ratings, 24 officers, one Wren, four civilians and five ladies who were to carry out welfare duties, the *Victorious* left Sydney. As the ship left her berth at Woolloomooloo she was given a very emotional send-off by a crowd of about 5,000, for in those days before the age of jet travel, such farewells were seen as final with few ever imagining that they might one day return to the land of their birth. Loaded on the flight deck were six Corsairs, six Fireflies, six Seafires and a Walrus, all of which were to be delivered to RNAS Trincomalee, but soon after leaving Sydney it became clear to Captain Annesley that there was not nearly enough space on the flight deck for recreational activities, and he obtained permission from the C-in-C British Pacific Fleet to dump the six Corsairs over the side. Perhaps the best account of this unique voyage, made by one of His Majesty's major warships, is told by Gwen West, whose husband Bill was serving in the *Indefatigable*: 'The mighty *Victorious* drew away from Woolloomooloo Wharf with us brides at the many vantage points and brightly coloured paper streamers trailing from the ship's side. Our families and friends, apprehensive and tearful, were waving goodbye from the wharf and gradually the Harbour Bridge receded into the distance. We were now subject to shipboard rules and we had all been issued with a booklet entitled, "HMS *Victorious* - A Guide To Civilian Passengers". Basically it said we were not to interrupt anyone in the course of their duties. The first shock came when we arrived in the dining hall, for after being subjected to quite severe rationing in Australia the variety and amount of food available was surprising. We all took our meals in the B hangar dining hall, with two sittings for breakfast, dinner, afternoon tea and supper.

The sleeping accommodation was brilliantly devised using cabin, recreation and hangar space to provide 700 berths. Bathroom facilities were adequate and with a laundry right forward on the port side we had no cause for complaint. Our washing could be hung out on lines provided in the forward lift well or, in bad weather, in C hangar, and ironing boards were provided in each dormitory. We also had access to a small shop in the wardroom lounge, but sweets, ice cream and soft drinks could be bought at the canteen on the flight deck.

Once outside Sydney Heads we really knew we were at sea. There was a dramatic change in the weather as we encountered mountainous seas whipped up by gale force winds, and as we crossed the Great Australian Bight those who were not good sailors spent a great deal of time in their bunks. The planes carried on deck had to be more firmly secured and the pounding of the heavy seas, together with the clangs and bangs, made sleep difficult. It was a tremendous relief when it all stopped on our arrival at Fremantle, and we sailed on the morning of 10 July to the strains of "Waltzing Matilda" played by a band on the wharf. Gradually the coastline slipped away into the distance of the Indian Ocean and it was our last view of Australia - very moving. Whilst at sea film shows were held daily and we had a library adjacent to the Chapel and the torpedo body room. There were materials for leather craft and for felt toy making, and we even had the luxury of a small hairdressing salon at the after end of the lounge; nothing had been overlooked. On 10 July there was the "Crossing the Line" ceremony when "King Neptune" and all his court arrived on board and it was an opportunity for the ratings to legally inflict aquatic punishment on their officers. A good day's fun and merriment for all.

We arrived at the British Naval Base in Trincomalee on the morning of 17 July, where the planes were offloaded and the ship was refuelled. The surrounding landscape looked very much like Sydney Harbour and it was a nice surprise to see such a beautiful waterway. In mid-afternoon we sailed in much calmer seas around the coastline of Ceylon, arriving at Colombo on the morning of 18 July. There the authorities were not taking any chances with our safety and escorts were provided from the ship's company for groups of us who wanted to go ashore. A launch took us from the ship to the wharf, and we caught our first glimpse of the "exotic east" with its own unique sights, smells and sounds, which was something I had only read about. The locals were well informed about our arrival and they also let us know that they were familiar with the names of our Australian cities, calling out to us "Mrs Sydney", "Mrs Melbourne" etc. We had a good time and found the shops, streets, markets and people most interesting, but we were glad to get back to the ship later in the day.'

On this occasion the Governor of Ceylon, Sir Henry Moore, and Lady Moore were embarked for the passage to the UK, and after taking on fresh provisions and water the *Victorious* left for Aden at 10am on Saturday 20 July. The story is taken up once again by Gwen West: 'Over the next few days the seas were calm and we saw ships passing at intervals which caused some excitement as we all relaxed on deck. We even held a sports day, with a lot of participants ending up with badly blistered feet from running on the scorching surface of the flight deck. In Aden a dance was held on the flight deck in a magical setting on a still, balmy night, and it was enjoyed by everybody. For the next two days we were in the Red Sea and we all felt the effects of the intense heat. It was too hot to do anything but sit around, although the ship's company had to keep going. As we steamed north through the Suez Canal the sight of an aircraft carrier full of females lining the flight deck seemed to give the local people quite a surprise as we could see their reaction to the sight of the ship - perhaps they thought the *Victorious* had been taken over by us. The memory of this giant aircraft carrier steaming through the Suez Canal will remain with me forever, but the journey ended in the evening when we arrived at Port Said. On the following day we went ashore and we were surprised to find that the Egyptians were quite hostile. There were also a lot of Italian sailors in the town and we witnessed quite a few ugly brawls, so all-in-all the atmosphere was not very congenial. However, we were able to shop and eat in the restaurants, but it was nice to return to the ship which had become our haven. The Mediterranean Sea was very blue and calm with ships passing on the regular trade route and over the next few days we enjoyed the beauty of the sea. Our arrival at Gibraltar was exciting and we were able to go ashore and explore, for the first time feeling quite safe as we walked round the streets. We left Gibraltar early next day on the final leg across the Atlantic Ocean, which lived up to its reputation of rough seas, and on Monday 5 August we spent the day packing our luggage and getting ready to leave the ship. On 6 August we slowed down to eight knots as we were ahead of schedule, and we anchored in Plymouth Sound at 6pm. There the Lord Mayor of Plymouth and an Admiral came on board to welcome us to Britain, and to the historic city of Plymouth - we all appreciated the gesture.'

Next day, at 12.30pm, the *Victorious* weighed anchor and on a fine summer's day she steamed up-harbour, passing between the West Hoe and Drake's Island, through the Narrows and into the Hamoaze, then past the Torpoint Ferry to berth alongside 6 & 7 wharves of Devonport Dockyard. For the 'war brides' it was certainly a far more impressive introduction to their new country than that afforded today by Heathrow and Gatwick Airports and Gwen West recalls her final hours aboard the ship which had been home for five weeks: 'Excitement was building as we stayed our last night aboard and everybody was up early for breakfast. We made our final departure from HMS *Victorious* at 7am on Thursday 8 August to be reunited

The *Victorious* at Bighi Bay, Grand Harbour, on Monday 11 November 1946, on her final trooping voyage. Note the deckhouse just forward of the island, which was a canteen.
(M. Cassar)

with the reason for our journey - our husbands. How fortunate we were to be given so much help and consideration by the ship's company on the wonderful voyage, for all the arrangements made for our comfort and, most of all, for having the privilege to travel with the Royal Navy in the aircraft carrier, HMS *Victorious*.'

Turning to the lighter side of the voyage, a number of 'brides' were most helpful and undertook all sorts of work from typing and clerical work in the enquiry office, to the 'brides' painting party' in their navy blue overalls. Included in the entertainment programme was the 'Prettiest Leg & Ankle' competition, and on sports day there was a 'Rescue the Maiden' race. One of the most enjoyable evenings was the fancy dress ball where the talent for devising original costumes was remarkable, and there were all sorts of handicraft classes. The entertainment programme was concluded just before the ship reached Gibraltar, when Lady Moore presented the prizes. This was followed by a cabaret evening at which considerable talent was shown by the ship's company and by the 'brides'. Perhaps the sight which most filled the old sailors' hearts with horror were the clothes lines of 'smalls' adorning the lift wells. There was an amusing incident in the Suez Canal as the ship passed a miltary 'camp' where huge numbers of bronzed young men watched and waved to brides who were lining the flight deck. Both sides started gaily waving to each other, until the voice of Captain Annesley came over the tannoy announcing, 'The young men you are waving to are German prisoners.' Among the more unusual duties for Officers of the Watch were 'Chastity Rounds'. In his report of the voyage Captain Annesley described them thus: 'Rounds of all weather decks, galleries and gun positions were carried out frequently and at irregular periods after dark. All women had to be in their bunks by 11pm and the duty woman officer went round to see that no women were missing. The women's sleeping accommodation was all enclosed and could not be entered by members of the ship's company. All the men's mess decks were out of bounds to women at all times. These measures were the best that could be devised and although by no means perfect, at any rate, acted as a deterrent to bad behaviour and broke up many petting parties before they reached their logical conclusion.'

On her return to Devonport the *Victorious* commenced a two-month refit with a long spell in No 10 dry dock and it was not until 31 October 1946 that she was once again ready for sea and a trooping voyage to the Far East. As before she embarked a very mixed draft of passengers, including Wrens and Nursing Sisters, and left Devonport to steam by way of Gibraltar, Malta, Port Said, Aden, Colombo and Trincomalee, arriving alongside No 8 berth of Singapore Naval Base on Sunday 1 December 1946. After leaving Singapore the next day she crossed the South China Sea and moored in Hong Kong Harbour three days later when her main drafts for HM Ships *Adamant*, *Cockade*, *Bonaventure*, *Euryalus* and *Tamar* disembarked. One of those on board for the voyage was Mr J. D. Phillimore who had been drafted to the depot ship HMS *Bonaventure*. Here he recalls the time spent on board the *Victorious*: 'I will always remember that voyage east because I was fortunate in that there were a number of friends with me and we were able to stick together. I can remember that there was a Royal Marines Band on board and when they held dances on the flight deck for the passengers, they provided the music. On one occasion I remember there was a Grand Ball in one of the hangars. Myself and other "drafties" helped to lay up tables at mealtimes which was, in fact, six times a day - two sittings each for breakfast, lunch and dinner. However, this unusual duty had its compensations as the passengers used to have a collection for us on each table and so we were able to supplement our pay. We carried out this unusual duty every day until we reached Hong Kong, and after I left the *Victorious* I never saw her again.'

After leaving Hong Kong with over 1,000 naval officers and ratings embarked as passengers, she steamed home by the same route taken on the outward voyage. Christmas Day was spent at Colombo and on New Year's Eve she left Aden for Suez, arriving in Gibraltar on Thursday 9 January 1947. Two days later she left for Spithead, and early on Monday 13 January, whilst steaming in very heavy seas off Ushant, she received a distress call from a French steamer, *Payer Quartier*, whose cargo had shifted in the heavy seas causing the ship to take on a dangerous list. Although Captain Annesley altered course towards the vessel, other ships were quickly on the scene and so the *Victorious* was not needed and was able to resume her voyage. Next morning she anchored at Spithead where the Portsmouth and Chatham ratings disembarked, and after making an overnight passage she moored to No 9 buoy in the Hamoaze, Devonport, at 12.30pm on Wednesday 15 January where all her passengers disembarked. With this voyage to the Far East completed it was, in fact, the end of the *Victorious*' career as a troopship.

After moving alongside No 6 & 7 berths at Devonport, work began on unloading the 'Food for Britain' cargo which she had brought back from the Far East, and deammunitioning. By early February the ship's company consisted of little more than a care and maintenance party, and on Tuesday 4 February the carrier was towed out to the Capital Ship Trot in the Hamoaze where she was secured to No 2 buoy. During the tow she was involved in a minor collision with the aircraft repair ship, HMS *Unicorn*, but only superficial damage was caused to the two ships. For four months the *Victorious* remained laid up at Devonport, but on Wednesday 25 June, after having been inspected by the C-in-C Plymouth, she left Devonport to steam under her own power to Portsmouth. After spending four hours

The *Victorious* in 1948 when she was commissioned as a training ship. *(Wright & Logan)*

anchored at Spithead she was manoeuvred into D Lock at Portsmouth Dockyard where she remained until late July. On Friday 11 July 1947 Captain J. C. Annesley relinquished command of the ship, handing over temporarily to Commander A. H. Hunt RN, who had transferred from the battleship HMS *Nelson*, and 17 days later the *Victorious* was secured in Admiralty Floating Dock 11, where she remained high and dry until the end of August whilst work was carried out to convert her for yet another new role.

The summer of 1947 was notably hot and sunny and the *Victorious'* much reduced ship's company enjoyed a relaxed refit routine with 'Hands to Bathe' piped each afternoon and evening without fail. It was not long before the future role of the carrier became known when it was announced that HMS *Nelson*, the flagship of the Home Fleet Battle Training Squadron, was to be relieved by the *Victorious*, with the elderly battleship destined for a short spell in the Reserve Fleet before being broken up for scrap. On Friday 19 September, AFD 11 was flooded and the *Victorious* was towed across the harbour and berthed again in D Lock to prepare for her new role.

The Home Fleet Training Squadron had been formed in July 1946 with the *Nelson* and the *Howe* (*Anson* joined later in the year), and it was based at Portland to provide, '...a new development in training methods by which recruits will now receive shipboard training much earlier in their careers.' Those who qualified for this training were Special Service Seamen who volunteered for seven years with the fleet and five as reservists. The training included elementary instruction in seamanship, gunnery and torpedoes, and the men would form half of the Royal Navy's peacetime strength. The three battleships were manned with reduced crews in order to ease the accommodation position ashore and, more importantly, in order to keep the battleships serviceable and in commission without having to use full complements of trained men. For most of the time the ships swung round buoys at Portland, but occasionally they made courtesy calls at more exciting ports. They could not, however, make prolonged voyages as there were not sufficient engine room ratings for long periods of watchkeeping. With the demise of the battleship and the rise of the aircraft carrier as the Navy's capital ship it was only natural that one of the Illustrious-class aircraft carriers should join the Training Squadron.

On Monday 22 September 1947, with her paying-off pennant flying proudly, HMS *Nelson* steamed out of Portland Harbour and on the following day she was manoeuvred into C Lock in Portsmouth Dockyard adjacent to the *Victorious* where the transfer of stores began. At 8.30am on Wednesday 1 October all the ratings who were under training left the *Nelson* and embarked in the *Victorious*. They were followed shortly afterwards by Captain E. B. K. Stevens DSO* RN, the commanding officer of HMS *Nelson*, approximately 20 officers and the instructional staff, and at 9am that morning Captain Stevens assumed command of the *Victorious*, now commissioned into the Home Fleet Training Squadron. Captain Hugh Owen, who as a young sub lieutenant was the captain's secretary, remembers Captain Stevens as, 'a great character, with a monocle, formidable to look at but very charming with a great sense of humour. He had spent most of his life in destroyers, including much of the war, and we all held him in great respect.'

During her spell in the floating dock the passenger accommodation in the hangar was converted into classrooms, gymnasiums, dormitories and mess decks. The space in the hangar which had served as a large recreation area and had been used for all sorts of occasions such as fancy dress parades and the 'Prettiest Leg & Ankle' competition became a parade ground for inclement weather conditions. In the hangar the 12 classrooms, with a space between each, were arranged in two blocks of six on either side of a central gangway. Part of the recreational space on the forecastle deck was converted into lecture rooms, with the provision of a library and a quiet room as well. There were few reminders of the wartime *Victorious*, although Hugh Owen remembers one such example: 'Deep down in the after end of the ship there was a row of about ten officers' heads, each of which had a girl's name painted on the inside of its respective door. Sarah was one of them and I think Sally was another. Apparently during the war they also had silhouettes of Japanese aircraft painted on them together with details of their armament and their capabilities - so that instead of thinking of "sweet nothings" while sitting on the throne the aircrew officers could study details of their enemies. Although the posters had gone, the names were still there as an unusual memento of her service in the Pacific theatre of war.'

The *Victorious* left Portsmouth during the afternoon of Tuesday 14 October 1947 to steam down the Channel for Torbay and for shakedown exercises. That evening she anchored off Torquay and for the next five days she carried out daily exercises at sea, returning to Torbay each evening. On 20 October she was joined by the *Anson* and *Howe* and after a four-day break at anchor in Torbay, all three ships sailed for Portland on Monday 27 October and that is where the *Victorious* spent the rest of the year.

On a typical weekday the Men under Training would start the day at 6am, and at 8.30am they would fall in by Divisions before going to their classrooms for instruction. At midday they had a 40-minute meal break, before heading ashore for an hour of sports after which they returned to the classrooms until 4pm when they secured for the day. On most days leave was granted from 4.30pm to 11pm and woe betide anyone who was adrift. It was at 8.30am on Wednesday 18 February 1948 that the *Victorious* put to sea again to take part in three days of anti-submarine exercises with the *Anson* and *Howe*, and this was followed on Monday 23 February by the Admiral's Harbour Inspection. At that time the Flag Officer Training Squadron was Rear-Admiral P. K. Enright who, perhaps better than anyone, understood the life of the new recruits, for he had joined the Navy as a Boy Seaman in the early years of the century and he was the first man from the lower deck to reach Flag rank. Hugh Owen recalls that he was, 'a wonderful man and a great character'. Following the hectic week of the Harbour Inspection, routine on board settled down again as the *Victorious* swung round her buoy and became a very familiar sight in Portland Harbour.

On Tuesday 11 May 1948 the unfamiliar vibration of the carrier's engines in motion was felt again on board as she made an early start at 5.30am for the island of Guernsey to take part in the third anniversary celebrations of their liberation from the German wartime occupation. It was 7.14am on 9 May 1945 when the Guernsey Fortress Commander, Major General Heine, surrendered unconditionally, and on board the *Victorious* now was Major H. W. Le Patourel VC, who had left the bank where he worked as a clerk to join the Royal Guernsey Militia, before transferring to the Hampshire Regiment. He was Guernsey's only bearer of a Victoria Cross. Also embarked were the Royal Marines and the Hampshire Regiment's Bands, and during the celebrations at St Peter Port a guard of 100 Men under Training, together with the Royal Marines Band, performed a Ceremonial Sunset. At 11.45am on Thursday 13 May an official cocktail party was held on the quarterdeck to which the Lieutenant Governor and C-in-C Guernsey, Lieutenant General Sir Phillip Neame VC, was invited and he raised a few eyebrows when he arrived on board, to the accompaniment of a 17-gun salute, wearing khaki battledress. However, he was very impressed by the Guard of Men under Training. Following the ceremonies the *Victorious* left St Peter Port on the afternoon of Friday 14 May, her departure having been delayed by fog, and by 9pm that same evening she was safely back on her buoy in Portland Harbour.

Eleven days after arriving back at Portland there was a change of command when Captain N. V. Dickinson DSO[*] DSC RN relieved Captain Stevens. He had joined the Royal Navy in 1915 and had served as a midshipman in the battleship HMS *Royal Sovereign*. He was certainly no stranger to naval training, for in 1934 he had served as the First Lieutenant of the Boys' Training Establishment at Shotley Gate, HMS *Ganges*. During the Second World War he had served with distinction and he had been awarded the DSO and had been Mentioned in Dispatches three times. He had served on Atlantic convoys in 1940, and in 'Operation Torch' and the Sicily landings in 1942 and 1943. In 1944 he was appointed as the Senior Naval Officer Northern Adriatic, where he earned the nickname, 'Baron Dickinson of the Dalmation Sea', and he had come to the *Victorious* from Berlin where he had been head of the Naval Branch. Two days later a Royal Marines detachment joined the ship and the following day the *Victorious* put to sea for six hours to take part in manoeuvres in Weymouth Bay. In July there were three more days of exercises at sea, and on Tuesday 27 July she steamed down the coast to Torbay where she anchored and took part in the highlight of the year - the Olympic Games. It had been 12 years and two abandoned Olympic Games since Hitler's propaganda spectacular in Berlin and despite the fact that they were dubbed the 'Austerity Olympics', and the fact that three major sporting nations, Germany, Japan and the Soviet Union, were absent, they were largely successful. In those days there were not the resources to construct new stadiums and so traditional venues were used, with rowing at Henley, shooting at Bisley and sailing at Torbay and Cowes. The main athletics arena was Wembley Stadium while the nearby Empire Pool hosted the swimming events. In Torbay the *Victorious* and the battleships *Anson*, *Howe* and *King George V* (flying the flag of the C-in-C Plymouth), provided a substantial naval presence, and on Monday 2 August the ships were dressed overall for the opening ceremony of the Games. At 8.45am that morning the Norwegian royal yacht, *Norge*, arrived in Torbay to the thundering of a 21-gun royal salute, and Crown Prince Olaf of Norway landed to light the Olympic flame at Torre Abbey, just off the seafront at Torquay. That evening an official cocktail party was held on the *Anson's* quarterdeck for 700 guests, including all the Olympic sailors, and next day the yacht racing started in earnest. Although the British 'Dragon' and '6-metre' crews did not win any medals (the classes being won by Norway and the USA), Hugh Owen recalls crews arriving on board the *Victorious* on the last day of racing just in time for an impromptu supper in the wardroom. In the event one of only three British medals was won by the yachtsmen when the crew of the 'Swallow' class won a gold in their event, and on Friday 13 August, the last day of the yachting events, the massed bands of the Royal Marines played the respective national anthems of each winning country as medals were presented in the closing ceremony at Torre Abbey Gardens. This was followed by the extinguishing of the Olympic flame and the following day the *Victorious* returned to Portland Harbour.

In September 1948 the Training Squadron said goodbye to Rear-Admiral Enright, who was succeeded by Rear-Admiral E. W. Anstice, and in October the *Victorious*

A final view of the 'old' *Victorious* as she lay alongside Middle Slip Jetty, Portsmouth, just a few weeks before she was paid off and was taken over by the dockyard for her long modernization refit. *(Maritime Photo Library)*

visited Spithead, Torbay and Falmouth, with the last two months of the year spent in Portland Harbour again.

On Tuesday 18 January 1949 the *Victorious* left Portland to steam north through heavy seas to Rosyth for a refit, and after sheltering in Belfast Lough from Wednesday 19 January to Friday 21 January, following the collapse of the forward lift, she arrived in Rosyth on Saturday 22 January. The refit, which included a lengthy dry docking, lasted until Friday 8 April 1949, and three days later she was again secured to her buoy in Portland Harbour. In May and June she made further visits to Torbay and in early July she anchored off Penzance before taking part in a NATO exercise, 'Verity', in the Bay of Biscay. Also taking part was the *Implacable*, which was the flagship of the Home Fleet, the battleship *Anson*, and the French submarines *Mille* and *Le Triomphant*. It was a rare chance for the Training Squadron to get in some sea-time and to be in the company of such a large collection of warships. However, by 8 July she was back in Portland Harbour and seven days later, on Friday 15 July, there was another change of command when Captain J. A. Grindle CBE RN took over from Captain Dickinson. In September 1949 the carrier visited Penzance and in the following month Plymouth, where she spent three days moored in the Sound.

In October 1949 came the news that the *Victorious* would be paying off in the early summer of 1950 and that the ship would be undergoing a major modernization refit, but there were still four months with the Training Squadron and on Wednesday 2 November 1949, flying the flag of Rear-Admiral E. M. Evans-Lombe, Flag Officer Training Squadron, the *Victorious* left Portland in company with the *Anson* to take part in Home Fleet convoy and anti-submarine exercises in the Irish Sea. The *Implacable* was the front-line aircraft carrier with 702 Squadron's Sea Vampire jets embarked, while the *Victorious*, on the other hand, assumed the humble role of a large merchant vessel in the convoy. It was the *Anson's* final fling as an operational ship for, on conclusion of the exercise on 12 November, her place was taken by HMS *Vanguard* and the *Anson* was laid up in Gareloch. That Christmas the mess decks were cold and bleak and on Saturday 28 January 1950 the *Victorious* left Portland for Gibraltar for her final exercises with the Training Squadron. She steamed south in company with the *Vanguard* and the *Implacable* and after carrying out fleet manoeuvres en route she secured alongside the detached mole on the morning of Thursday 2 February. During her seven days in Gibraltar she was visited by the C-in-C Home Fleet, Admiral Sir Philip Vian, who had last been on board the *Victorious* when she formed part of his command in the Pacific Fleet. At 8.30am on Friday 10 February the *Victorious* left Gibraltar and, together with the *Vanguard*, she set course for Portland where she arrived four days later.

After just six days at Portland the *Victorious* left for Devonport during the afternoon of Monday 20 February, and the following morning she arrived alongside No 6 berth where stores were disembarked. Finally, on the morning of Tuesday 7 March 1950 she left Plymouth and that evening she anchored at Spithead, where she remained until 2.30pm the next day, when, with her paying-off pennant flying, she steamed up-harbour to berth alongside South Railway Jetty. Eight days later Ceremonial Divisions were held on the flight deck before most of the Men under Training transferred to the *Indefatigable*, now commissioned into the Training Squadron, and destoring and deammunitioning began in earnest. On Friday 24 March Captain Grindle handed over command to his Executive Officer, Commander Clive Gwinner DSO RN, and at noon on Friday 30 June 1950 HMS *Victorious* paid off and was taken over by Portsmouth Dockyard for a major reconstruction which would last for seven and a half years.

On the morning of Monday 3 February 1958 the *Victorious* put to sea for the first time in almost eight years. Here she is seen leaving Portsmouth Harbour. In the background are Fort Blockhouse, HMS *Dolphin* and the training ship *Foudroyant*. *(Fleet Air Arm Museum)*

This stern view of the *Victorious* in the Solent shows her new, fully-angled flight deck to its best advantage. *(Fleet Air Arm Museum)*

The *Victorious* undergoing sea trials. *(Fleet Air Arm Museum)*

Chapter Eleven

Almost A New Ship

Following the end of the Second World War developments in naval aviation were both rapid and revolutionary, with the first major event taking place on Monday 3 December 1945 in the Channel when Lt-Cdr E. B. Brown landed a jet-propelled aircraft on the flight deck of an aircraft carrier at sea for the first time. With much heavier and faster aircraft coming in to land at higher speeds, the axial flight deck with its two barriers separating the incoming aircraft from those parked forward of the island was clearly unsatisfactory. The solution, which was pioneered by Captain Dennis Cambell RN, was the angled flight deck which removed the deck park from the path of landing aircraft. Another problem when operating heavy jet aircraft was the fact that hydraulically powered catapults were unable to provide the thrust required, and the answer to this was the steam catapult. Another important innovation was the mirror landing sight, which was the idea of Lt-Cdr Nicholas Goodhart and which allowed a pilot to monitor his own landing thus making the need for a landing signals officer and his 'bats' obsolete. All these major improvements would be incorporated into the 'new' *Victorious*, but it was to be a long and tortuous process.

In October 1950 the modernization of the *Victorious* started in earnest and it was the largest task of its kind ever undertaken in a Royal Dockyard. The extent of the work can be assessed by the fact that at one stage some 2,300 men, mainly from the Dockyard's Construction Department, were involved both on board the ship and in workshops ashore. However, before the reconstruction work could be commenced nearly 15,000 tons of armour plating, machinery, electrical gear and general fittings had

Full-power trials create quite a bow wave, and a wake. *(Fleet Air Arm Museum)*

105

to be removed. During the reconstruction the underwater beam of the ship was increased by eight feet and the depth from keel to the flight deck level was increased by four feet. The *Victorious* was completely rebuilt above the hangar deck, and the upper gallery deck (immediately below the flight deck) was unique in a British carrier in that it was continuous for the full length and breadth of the ship. The 775ft-long flight deck was raised by four feet and to achieve its fully angled deck it was extended outwards on the port side for 41 feet over a length of 120 feet. It was strong enough to take the heaviest naval aircraft which could then be envisaged, including the Blackburn NA 39 (Buccaneer), and two parallel-track 145-ft steam catapults were fitted forward with aircraft positioners and jet blast deflectors. Steam for the catapults was provided from wing boiler rooms which were situated forward on No 6 deck. The arresting gear comprised four wires with an average span of 80 feet, and a single nylon emergency barrier was fitted. Deck landing mirror sights were fitted port and starboard and the angled deck necessitated the port sight being mounted well outboard on its own sponson. As before the two hydro-pneumatically operated centreline lifts would handle the required working load, but the ship's most striking feature was the $8^{3/4}°$ fully angled flight deck, the enormous overhang being supported by a large sponson which was bracketed into the carrier's hull structure. The overhang represented a compromise between the claims for the most efficient flight deck layout and the need to prevent the risk of damage in a heavy seaway. The island superstructure, which was small by the standards of the day for the size of the ship, was kept so in order to afford the maximum available area of flight deck, and the siting of the two-tier operations room and radar display room below flight deck level also enabled the size of the island to be kept to a minimum. The main visual feature of the superstructure was the huge Type 984 radar with its 'searchlight' scanner which occupied the forward position. It was the first of its size to be mounted in a warship and in order to accommodate it the lattice mast was fitted to the after end of the island, abaft the funnel.

The main hangar was divided into two sections by an asbestos fire curtain, while the lift openings to the main hangar could be sealed off by hydraulically operated hangar doors. Stowage was provided for the more conventional aircraft ammunition such as bombs and torpedoes, and the bomb rooms were fitted with adjustable bins for the various sizes of weapon which were in service and under development, and two electrically operated platform bomb lifts served the hangar and flight deck. The after bomb lift also provided an alternative supply route for food from the galley to the wardroom in the event of a breakdown of the food lift. During the reconstruction new boilers were installed so that the additional demand from the new turbo-generators and the steam catapults could be met, and the main machinery could be operated by remote control from a central Machinery Control Room.

From the ship's company's point of view the most important alterations were made to the living accommodation on board amid the inevitable battles for space for the ever-mounting volume of technical equipment and the needs of the crew. In the final compromise all the officers were provided with single-berth cabins, which occupied less overall space than the old two-berth and dormitory cabins. All the mess decks were fitted with standee fold-up bunks for ratings and there were separate dining halls for junior and senior ratings. The main galley was equipped to produce three, four-course meals a day for 2,000 men, and the bakery could provide 3,000 loaves of bread daily. The canteen was the largest to be installed in any British warship at that time and it sold everything from soap powder to ice cream.

In all the cost of the modernization was almost £15 million, and there were many who thought that this money would be better spent building a new ship. In retrospect their argument has some merits for, less than ten years after the modernization, it was certainly easier for the politicians to send a rather elderly carrier prematurely to the scrapyard than it would have been if they had been faced with a new and relatively trouble-free vessel. However, the rebuilding of the *Victorious* was rather a haphazard affair as it was continually prolonged in order that the latest technology could be utilized. Originally the modernization was to have been completed by the summer of 1954, but in 1953 the angled deck, the steam catapults and the new boilers were added to the work schedule, and in the following year alterations to the hangar design were approved. In 1955 and 1956 further amendments to the specifications were made and it looked like the *Victorious* would become a permanent fixture in the landscape of Portsmouth Dockyard.

Statistically the modernization of the *Victorious* makes impressive reading, with 800 miles of electric cables, 10,000 lighting points, ten miles of ventilation trunking and 17,000 square yards of linoleum having been fitted into the ship, while in C Lock some 400,000 feet of tubular scaffolding was used in staging up the ship. One of the many problems which had to be tackled was the serious shortage of shipwrights at Portsmouth Dockyard, and consequently Devonport Dockyard was delegated the prefabrication of No 2 deck and the new island supersructure with the 30-ton sections being transported to Portsmouth by sea.

On Wednesday 1 January 1958, with most of the structural work and the fitting out having been completed, the *Victorious* was towed from C Lock to Pitch House Jetty, a move which necessitated the removal of the dockside crane to allow the clearance of the port side overhang of the new angled deck. Two days later she was visited by the C-

Tilt trials in the Solent. First to port...

...and then to starboard. *(Fleet Air Arm Museum)*

The *Victorious* on 2 October 1958, off Gibraltar, with Scimitars and Sea Venoms on deck.
(Fleet Air Arm Museum)

in-C Portsmouth, Admiral Sir Guy Grantham, and at 8am on Tuesday 14 January 1958 the 500-strong commissioning parties of the ship's company marched from the Royal Naval Barracks via Lion Terrace, Park Road and The Hard to the dockyard. The contingent was led by the Royal Marines Band and it was said to be the largest commissioning party to have left the barracks since the end of the Second World War. Some relatives and other early morning onlookers watched their departure from the barracks, and there were more at the Main Gate of the dockyard to see them enter. On their arrival at the ship they were met by Heads of Departments and one contemporary report observed: 'They must have noted with some satisfaction that among the contingent of petty officers were a large proportion of "stripeys" - holders of three good conduct badges, indicating two assets which will be of the greatest value in this virtually new carrier - experience and length of service.' At 9am that morning Colours and the Commissioning Pendant were hoisted, and at 2.15pm the Royal Marines detachment paraded three Guards of Honour and the band on the quarterdeck for visiting Flag Officers, and at 3pm the Commissioning Service was conducted in B hangar by the Chaplain of the Fleet with Captain C. P. Coke DSO RN reading the Commissioning Warrant to the assembled ship's company.

After a fortnight of preparation and settling down, during which further drafts of the ship's company joined her, at 11am on Monday 3 February 1958 the *Victorious* slipped her moorings and put to sea for the first time in almost eight years. Hundreds of spectators stationed themselves at the Sally Port and along Southsea seafront to watch the carrier steam out of the harbour and into the Solent, and the maze of little streets at Old Portsmouth was congested with cars and bicycles. Her departure was a real 'naval occasion' for Portsmouth and it even overshadowed that of HMY *Britannia*, which left harbour ten minutes earlier. At the forward end of the flight deck were two Whirlwind helicopters and aft was a Fairey Gannet. As is customary, the ship's company manned the flight deck, but a very unusual sight was a large contingent of dockyard technicians and workmen, who were to accompany the ship throughout her sea trials, forming a neat square on the forward lift. They had the responsibility of seeing that the ship's machinery and other equipment was running properly before the carrier was handed over to the Navy. The *Victorious* was preceded by the tug *Capable* which, as though sensing that the 'Old Lady' had not been to sea for some years, kept the tow rope in place until they were well out of harbour.

The carrier's preliminary sea trials lasted for the whole of February and they were marred by a fatal accident on the third day, Wednesday 5 February, while the ship was steaming in the Channel off Portland and landing on one of the helicopters of 701C Flight. As the machine touched down on the flight deck part of its tail rotor broke off and hit Naval Airman Russell Ward, killing him instantly. Next day the ship hove to off Plymouth breakwater and the body of the unfortunate rating was landed by MFV. Forty-eight hours later the carrier was back at Spithead for three days of heeling trials which caused a certain amount of amusement and the occasional embarrassment to some. On Saturday 15 February the *Victorious* left the Solent to make a choppy passage north through the Irish Sea to the Clyde in company with the destroyer *Contest*. During the seven days in the area speed trials were carried out on the Arran measured mile, and on Sunday 23 February she set course for the south coast once again to carry out further trials. On the last day of February, whilst the ship was in Lyme Bay, Lt-Cdr P. Lamb RN of No 700 Carrier Trials Unit, RNAS Ford, carried out the first 'bolters' of the commission when he made several neatly executed touch-and-go circuits in a new Supermarine Scimitar. Next day, at 8am, the *Victorious* secured alongside Portsmouth's Middle Slip Jetty for a further period in dockyard hands.

During her stay in Portsmouth the ship was visited by a number of interested VIPs, including a Turkish Parliamentary delegation, the US Naval Attaché and the First Sea Lord, all eager to see the 'new' aircraft carrier. On Tuesday 11 March the Chaplain of the Fleet conducted a service of dedication for the ship's chapel, which was a haven of tranquility in a large, busy and noisy vessel. Built by the dockyard it was designed so that one could slip easily from the hustle and bustle of a modern warship into the peace and quiet of the church, which was panelled in light wood and contained a stained-glass window and candlesticks from the battleship HMS *Nelson*. During the Easter holiday of 5 to 7 April the *Victorious* was opened for Navy Days and over the three-day period more than 18,000 visitors took the opportunity to look over the carrier. Meanwhile, at RNAS Ford in Sussex, service trials of the Navy's newest, fastest, heaviest and noisiest aircraft, the Supermarine Scimitar, by a flight of 700 Squadron were nearing completion. The Scimitar was to enter squadron service at Lossiemouth on 3 June 1958 and would be embarked in the *Victorious* in the autumn of that year. It was designed to replace the Sea Hawk as an interceptor, but it was also capable of carrying both nuclear or conventional weapons and of being refuelled in the air. To cope with the noise of the aircraft the flight deck parties were issued with a new type of earmuff. Back on board the *Victorious* the new steam catapults were undergoing deadload trials with each catapult carrying out 12 launchings a day under the watchful eye of a team of experts from RAE Bedford, Officers from the Department of the Admiralty Engineer-in-Chief and from the Flight Deck Machinery Trials and Training Unit.

After each watch on board had taken their leave periods, members of their families embarked at 7am on Wednesday

A Fairey Gannet landing on. (*Fleet Air Arm Museum*)

28 May for a day at sea in the Channel at the end of which the *Victorious* anchored in St Helen's Bay off the Isle of Wight and the visitors disembarked in tugs. During the remainder of the month final sea trials were carried out in the Channel and finally, on Monday 23 June, Fairey Gannet 504 flown by Commander S. J. A. Richardson RN performed the first landing on the new flight deck. Later that same day he also made the first catapult launch and over the following three days everyone became immune to Sea Venoms, Sea Hawks and Gannets, and the Scimitars of No 700 Carrier Trials Unit created a new sensation when they landed on board for deck landing trials. They were complemented by the prototype De Havilland 110, the Sea Vixen, from Boscombe Down flown by Lt-Cdr D. P. Norman, which also made a deck landing. At one stage it was said that the flight deck resembled the short runway at Heathrow Airport. Then came a long weekend break at the end of the month when, on the morning of Friday 27 June, the *Victorious* secured alongside the Quai de Joannes Couvert at Le Havre for a three-day visit to the French port. This was followed by further intensive trials in the Channel before the carrier returned to Portsmouth and moored alongside Pitch House Jetty on the morning of Friday 11 July.

The first weekend in August was a public holiday and the ship was visited by Prince Michael of Kent then by over 20,000 members of the public on Navy Days. At the end of the month the *Victorious* left Portsmouth for a further period of flying exercises which were completed on Friday 5 September when she returned once again to Pitch House Jetty in Portsmouth Dockyard. During the 20-day stay the visitors' book recorded more eminent names including Admiral Sir Manley Power and the C-in-C Portsmouth, Admiral Sir Guy Grantham. On Tuesday 23 September the ground parties of 893 Squadron (Sea Venom) and 849 B Flight joined the ship which indicated that the commission was starting in earnest, and the next day the Whirlwind helicopters of 824 Squadron, under the command of Lt-Cdr J. Trevis RN, were landed on, although they had officially joined the ship on Wednesday 27 August.

At 11.10am on Thursday 25 September 1958 the guard and band paraded for leaving harbour and the flight deck was manned as the *Victorious* slipped her moorings and left Portsmouth to embark her squadrons and to carry out her work-up before joining the Mediterranean Fleet. Several hundred people lined the foreshore at Sally Port and Old

Portsmouth as the ship left harbour, firing a salute which was answered by the saluting battery at Fort Blockhouse. As she steamed into the Solent she presented a fine contrast to the outward bound liner *Queen Mary*, which was en route to New York. However, for the *Victorious* it was to be a very sad day.

At 1.42pm that afternoon, when the ship was in the Channel, 15 miles off Dunnose Head, Isle of Wight, she turned into the wind and increased speed in order to land on the Scimitars of 803 Squadron. One eyewitness describes the events which followed at 2pm: 'Two Scimitars arrived overhead and each made three dummy landings. The first then came in and appeared to make a perfect landing as his trailing hook caught an arrester wire. Flight deck ratings scattered as the wire shot across the deck, and the aircraft went over the side into the sea and floated on the surface. A helicopter was standing by and it moved in with a man dangling at the end of a rope ladder. Then suddenly the aircraft sank.' The Scimitars had flown from RNAS Lossiemouth via Yeovilton in pairs and, in the full glare of press publicity, this tragic accident had befallen Commander J. D. Russell RN, the commanding officer of the newly formed 803 Squadron, after he had made what appeared to have been a perfect landing. In the event it was subsequently found that the cause of the accident was a faulty valve in the arrester gear which had led to a malfunction, and this had been followed by a series of unfortunate events which had led to the death of the pilot. A naval witness at the Coroner's Inquest gave the following evidence: 'Commander Russell was not able to get out of the cockpit of the Scimitar because of a chain of circumstances rather than of an individual one. First he tried to jettison his hood, but had difficulty in operating the release handle. This would constitute a time-waster while he sorted out the alternative means of getting the hood open. Secondly Commander Russell had difficulty with his oxygen apparatus and he would have wasted further time taking off his helmet and oxygen mask. He tried to open his hood manually, which was not easy; he partially succeeded but it closed again. He did successfully open it after the aircraft had submerged, but at that point he was trapped in the cockpit by his leg restraint - a method of restraining the legs if the ejection seat is used - and also by the dinghy lanyard connexion between his life jacket and his dinghy.'

As soon as the Scimitar hit the water the planeguard helicopter, flown by Lt R. A. Duxbury RN, and crewed by Lt G. R. Fyleman and Ldg Seaman Brown, made a very brave but unsuccessful rescue attempt. For over two hours a thorough, but fruitless, search was made of the area and, in fact, it was late October before the wreck of the Scimitar and the body of Commander Russell were recovered by the boom defence vessel HMS *Barfoss*. It was 4.30pm on the afternoon of 25 September before flying was resumed and the remaining aircraft of the squadron were safely embarked, now under the command of Lt-Cdr G. R. Higgs RN. They were followed by the Sea Venoms of 893 Squadron, commanded by Lt-Cdr E. V. H. Manuel RN and the Skyraiders of 849 B Flight under the command of Lt-Cdr B. H. Stock RN.

Two days later, at 12.10pm on Saturday 27 September, whilst the ship was carrying out flying exercises in Lyme Bay, Whirlwind helicopter XL 848 crashed into the sea about one mile off the port bow, but, fortunately, the crew were recovered by the SAR helicopter. Although all three crewmen were injured they were quickly transferred to hospital and they made a complete recovery, but the three days of flying had provided a stark reminder of the hazards of naval aviation. At 1.45pm on Sunday 28 September the *Victorious* left Weymouth Bay and set course for Gibraltar. On the following day the wind freshened whilst she was on passage through the Bay of Biscay, and by nightfall it was blowing at gale force. During the night wind speeds of 72 knots were registered and four life-rafts were carried away. Speed was reduced to eight knots, but despite this the heaving seas caused considerable damage to the starboard catwalk and down below the forward dining hall was more like a skating rink of tea and soup. Fortunately, the severe weather subsided later in the day and during the evening the ship passed Cape St Vincent. Next day, whilst within sight of Gibraltar, the Scimitars and Venoms were carrying out simulated strikes on the airfield at Gibraltar's North Front, when a signal was received from Sir Winston Churchill who had been cruising in his yacht *Christina*, and who was about to fly home from North Front. The signal stated that he would be delighted to see the ship's aircraft before his departure and the squadrons duly obliged with the airborne Scimitars beating up North Front as the 'great man' was about to board his plane. Apparently this pleased him greatly. At 9.10pm that evening, when the ship was off Europa Point, two ratings, AB D. Kennedy and M(E) P. M. Kennedy, who were brothers, were reported missing, together with one life-raft, and a search was instituted. While lookouts were being posted, one man, who was unaware that the starboard catwalk had been damaged in the gales, jumped into where it had once been from the flight deck and, instead of landing just a few feet down, he found himself falling into the sea. Fortunately, his cries for help were heard by several officers and men and all four quarterdeck lifebuoys were dropped simultaneously. The ship was steaming at 15 knots at the time but it was immediately turned round and the seaboat was lowered, enabling the unfortunate soul to be picked up just 11 minutes after he had fallen into the sea. On entering the sickbay after his ordeal his only comment was reported to be, 'I'm only in for two years. I don't want to make a habit of this.' In the search for the missing brothers the life-raft was sighted in the moonlight at 11.30pm about

On Friday 20 February 1959 the *Victorious* left Portsmouth for the Mediterranean. In this view she is passing the Round Tower. In the background can be seen the incomplete aircraft carrier *Leviathan* and, laid up, the Royal Navy's last battleship, HMS *Vanguard*. *(Fleet Air Arm Museum)*

HMS *Victorious* in the Firth of Forth on 31 May 1959, with the Forth Bridge in the background. *(Fleet Air Arm Museum)*

two and a half miles away and by 11.40pm the two absconders and their escape craft had been recovered. Apparently one of the two brothers was in trouble on board and he had persuaded the other to accompany him to what must have appeared the glamour of southern Spain. In the event they had launched a life-raft from the after end of the flight deck and had jumped after it, but luckily the incident did not end in tragedy, particularly for the unfortunate lookout.

At 7.45am on Friday 3 October the *Victorious* entered Gibraltar and secured alongside 47 berth on the South Mole for a three-day break with the opportunity to do some shopping or to take part in the numerous sporting fixtures which had been arranged. Early on the morning of Monday 6 October the carrier left Gibraltar for further flying exercises east of the colony, and three days later she moved along the coast for a day's flying off Oran. During the exercises a Scimitar was diverted to the French Air Force base at La Senia south of Oran after a failure of its hydraulic system due to damage sustained on take-off. Later in the day a Skyraider was sent to the French base with the Squadron AEO to assess the damage, and once this was known the *Victorious* herself closed Oran breakwater to put spare parts and a maintenance party ashore. Following the delay caused by this diversion the carrier arrived in Marsaxlokk Bay, Malta, during the late afternoon of Sunday 12 October, where three Whirlwinds which had been carried as cargo were unloaded, and next day she steamed into Grand Harbour where she secured to 11 buoy, next to HMS *Eagle*. During the ship's stay in the Mediterranean the *Victorious* played host to a film crew who, under the direction of Mr Harold Baim, made a film about the ship. Less popular was an Admiralty Work Study Unit who, it was rumoured, had been instrumental in making 'loafing' an even more skilled art than it had ever been. After only two days in Grand Harbour the *Victorious* sailed out for her first work-up period in the Mediterranean, but because of problems that had been experienced operating the Scimitars from the catapults, Venoms were used instead while the helicopters performed anti-submarine exercises. The period was enlivened by a night encounter exercise with the 3rd Destroyer Squadron (*Saintes*, *Camperdown* and *Armada*) during which the *Victorious* took on the role of a six-inch cruiser. During the exercises the carrier was visited by the Flag Officer Malta, Rear-Admiral Sir Charles Madden, and the FOAC, Vice-Admiral A. N. C. Bingley, both of whom embarked by Skyraider. The end of the exercise was marked by a ten-day visit to Toulon, whilst the squadrons flew off to Hyeres. During the visit hospitality was exchanged with the cruiser *De Grasse* and the aircraft carrier *Arromanches* (ex-HMS *Colossus*) and the *Bois Belleau* (ex-USS *Belleau Wood*), then on the morning of Monday 3 November the *Victorious* left the 'Cité Maritime' to take part in more exercises off Malta.

At 11.13am Whirlwind P suffered engine trouble and ditched some five miles from the ship but, fortunately, the crew were rescued by other helicopters which were taking part in the exercise. Four days later the planeguard helicopter suffered a similar mechanical problem, but the pilot was able to land on safely. As a result of these difficulties 824 Squadron were grounded, but following the loan of two Whirlwind Mk 3s from 848 Squadron at Hal Far they were able to resume flying. On Friday 14 November the ship paid a four-day visit to Messina during which tours were arranged to Gambaari and Taormina and an intrepid band of climbers made an unsuccessful attempt to scale Mount Etna. When the carrier was opened to visitors, so many people wanted to get on board that the local police were called to prevent crowds from charging the brow. After leaving Messina the *Victorious* joined the *Eagle* off Malta and for the first time since her modernization she operated with another carrier. On completion of these exercises on Friday 28 November the Venoms and Skyraiders were dispatched to Hal Far, while the ship entered Grand Harbour for ten days' self-maintenance. During a 13-day period at sea before Christmas the Venoms of 893 Squadron carried out successful firing trials of the Firestreak missiles on drone targets, and some simulated night attacks were made on the cruisers *Sheffield* and *Ulysses* which were steaming from Gibraltar to Malta. The night flying was completed by operating round the clock for some 58 hours, during which time small sorties of Skyraiders and Venoms carried out day and night exercises. On conclusion of the exercises the *Victorious* entered Grand Harbour on Monday 22 December for the Christmas break, which was sadly marred by the death of Steward R.J. Walker who drowned in the harbour on Christmas Day.

On Friday 2 January 1959 the *Victorious* left Malta and after a hectic five-day exercise with RAF Canberra bombers she set course for Gibraltar before heading home to Portsmouth. After flying off her Air Group, including the Scimitars, on the morning of Tuesday 13 January 1959, she anchored at Spithead that evening. Next day, on a cold and foggy afternoon, 1,100 relatives were embarked for the passage up-harbour and at 4.15pm on 14 January the *Victorious* secured alongside Pitch House Jetty. It was exactly a year to the day since she had commissioned.

The *Victorious* sailed again from Portsmouth on the morning of Friday 20 February, having been delayed for 24 hours by thick fog which had blanketed most of the south coast. Next day the squadrons were embarked safely and course was set for Gibraltar and the Mediterranean where, at 7.15am on Friday 27 February, she rendezvoused with the *Eagle* off Cartagena. After carrying out search and strike exercises the *Victorious* returned to Gibraltar where, after seven days alongside, she rejoined the *Eagle* and units of the Home Fleet and the French and Dutch Navies for 'Exercise

Dawn Breeze IV'. However, just before the start of the exercise the helicopters of 824 Squadron were grounded again owing to a mechanical fault and during Monday 16 March they were transferred to the *Eagle* by lighter, in exchange for six Venoms of 894 Squadron which were transferred from the *Eagle* in the same manner. The exercise itself was designed to test the NATO fleet's air defences and it included RAF 'V' bombers and Canberras, with *Eagle*, *Victorious* and *Centaur* forming a fast carrier striking group. During the exercise two Scimitars were diverted to the *Centaur* for an overnight stay, which made for a very crowded deck on board the light fleet carrier. The exercise finished on Sunday 22 March and all the ships returned to their home ports, with *Victorious* arriving in Portsmouth Dockyard's D Lock during the afternoon of Tuesday 24 March for dry docking and maintenance.

The *Victorious* sailed again on Monday 4 May 1959, and at 5.35pm, after re-embarking the Scimitars of 803 Squadron, she landed on three Sea Vixens of 77 Flight, commanded by Commander M. Petrie RN, for three weeks' carrier experience. After a short stop in Plymouth Sound where the FOAC, Vice-Admiral Charles Evans, hoisted his flag in the *Victorious*, the carrier sailed to embark her Venoms and Skyraiders. There then followed intensive flying exercises in the Channel which gave the Sea Vixens the chance to carry out plenty of deck landings so that after three days they were landing comfortably, although to the uninitiated they appeared to be the world's noisiest aircraft. On the weekend of Saturday 9 May there was a short break at Torquay and four days later the *Victorious* was in the North Sea to carry out air defence exercises with RAF Bomber Command and HMS *Centaur*. The *Victorious* conducted a search and strike exercise against *Centaur* which was returning from a 'jolly' at Copenhagen. A Skyraider detected her off the Norwegian coast and shadowed her whilst both Scimitars and Venoms, led by a Sea Vixen, carried out the air strikes. The combined exercises with *Centaur* continued in the Irish Sea with a break on Sunday 17 May when both ships anchored in Tremadoc Bay. Next day the two carriers were subjected to intense air attacks by aircraft from RNAS Brawdy, and at one stage it seemed that the whole approach to the Bristol

An unusual aerial view of the *Victorious* as she enters Oslo on the morning of Wednesday 17 June 1959.

(*Fleet Air Arm Museum*)

The *Victorious* arrives at Pier 90 in New York city on the afternoon of 30 July 1959. The French liner in the background is the SS *Liberté*.
(Fleet Air Arm Museum)

Channel was filled with aircraft. Next day the *Victorious* bade farewell to the *Centaur* which set course for Lisbon, while the former steamed for the Solent and anchored at Spithead on Tuesday 19 May.

During the latter half of May the *Victorious* carried out a number of 'Exercise Shop Window' programmes, demonstrations of the latest methods and equipment, for representatives from the Ministry of Supply, the Society of Aircraft Constructors and other interested bodies. The exercises were marred by fog and, at 3.21pm on Tuesday 26 May, by a tragic fatal accident when Naval Airman Thomas H. Craig was hit by a Sea Vixen as the aircraft was catapulted from the flight deck. On Friday 29 May the exercises were concluded and after flying off the Sea Vixens the *Victorious* set course for Rosyth where she arrived two days later. It was 4.10pm on Tuesday 9 June, after a 36-hour delay because of high winds, when the carrier left Rosyth to join the Home Fleet for a five-day air defence exercise with the RAF and Danish and Dutch forces, including the HMNS *Karel Doorman* (ex-HMS *Venerable*). During the exercise the two carriers cross-operated Venoms and Seahawks, and at 11am on Thursday 11 June HRH Prince Bernhardt of the Netherlands made a four-hour visit to the *Victorious*, arriving by helicopter. Two days later the *Victorious*, together with HMS *Tyne* and a number of destroyers, anchored at Arhus in Denmark for a pleasant 48 hours before steaming north to Oslo, where they arrived on the morning of Wednesday 17 June. During the afternoon of the next day King Olaf V of Norway visited the ship for just over an hour, during which time he walked round the vessel and was shown the cockpit of a Scimitar. The *Victorious* left Oslo on Saturday 20 June for Portsmouth and after carrying out full-power trials in the North Sea she arrived alongside Pitch House Jetty during the afternoon of Monday 22 June, to carry out essential maintenance and to prepare for the highlight of the commission - a visit to the United States of America.

After embarking press representatives the *Victorious* left Portsmouth with units of the 5th Frigate Squadron, including *Scarborough*, *Tenby* and *Salisbury*, and they set course for Norfolk, Virginia. The carrier was scheduled to pay a series of visits to US ports to give air defence demonstrations highlighting the new Type 984 'three dimensional' radar system, about which a great deal of interest had been shown in the USA. The electronic equipment installed in the *Victorious* had cost over £1 million and Admiral of the Fleet Lord Mountbatten had described the radar as the, '...finest set in the world. Nothing can touch our Type 984 that I know of.' During flying exercises on 6 July in the vicinity of the island of Laertes, a signal was received that a USAF B47, en route from Spain to America, had caught fire whilst air refuelling and the crew of three had baled out. A US supply ship and the destroyers USS *Camp* and USS *Kretchmer* were searching an area north-west of Flores. A British merchantman rescued the navigator of the aircraft and he was then transferred to the USS *Camp*. However, as he was needed back at his base, a helicopter from the *Victorious* collected the officer, together with a rating from the *Camp* who was suffering from appendicitis, and returned them to the carrier, from where a Skyraider flew them 280 miles to USAF Lajes. It was a good deed done on Independence Day. After some successful air defence exercises in the Western Atlantic against US Navy Skyhawks and Banshees, the *Victorious* arrived off the Chesapeke Light Vessel early on the morning of Friday 10 July and, leading the three frigates, she came alongside in the Norfolk Naval Base at 9.30am. It had been 16 years since her last visit and, once again, Anglo-US relations were cemented in many ways. After four very pleasant days enjoyed by everyone, the *Victorious* sailed on Tuesday 14 July for 'Exercise Riptide' with the US Navy's 2nd Fleet under the overall command of Vice-Admiral Smedberge, who was flying his flag in the USS *Northampton*. The US Navy provided two fast carriers for the exercise, *Saratoga* and *Essex*, with the USS *Wasp* as the ASW carrier, together with 35 destroyers, three oilers and a supply ship. The exercise lasted from 15 to 20 July and during this time the *Victorious* cross-operated US Navy Crusaders, Skyhawks, Skyrays, Demons and Tracker aircraft, sending her Scimitars to *Saratoga* and the Venoms and Skyraiders to both US carriers. During the course of the exercise Admirals were also exchanged, with Admiral Evans visiting the *Saratoga*, and no fewer than five US Admirals plus the US Ambassador to London embarking in *Victorious* to see a demonstration of the 984 radar. By Monday 20 July the *Victorious* had made her mark in the air defence phase of the exercise, and her aircraft had carried out day and night raids, with the Scimitars attacking targets in North Carolina. During the cross-deck operating a Demon FH3 required a tyre change and this was soon accomplished by a Skyraider. One of the USN Crusader F8Us was found to have a fractured launching box, and it was a very impressed pilot who was catapulted safely after the part had been replaced on board the *Victorious*. At the end of the exercise there was another short visit to Norfolk and this was followed by a four-day stay in Boston, Massachusetts. However, the entry into New York, which had been scheduled for the morning of Thursday 30 July, was delayed for eight hours because of an accident involving the liner *Queen Elizabeth*, whose berth the *Victorious* was to occupy. After a frustrating day anchored off the Ambrose Light Vessel the carrier finally secured alongside Pier 90 in New York City at 5.30pm. As with the other two US ports the local people took the *Victorious* to their hearts by inviting a good number of the ship's company into their homes and showing them their brash, loud, but friendly city. Sightseeing took in all the landmarks from the Empire State Building to Greenwich

The *Victorious* looking very smart in the Solent on 15 September 1959. She has Scimitars, Venoms, Gannets and a Dragonfly helicopter on deck. *(FotoFlite)*

Village, and the US Marines surprised the Royal Marines detachment by bringing their adopted 'pin-up' girl, Nancy, all the way from Boston to visit the ship. But, alas, all good things had to come to an end and on the morning of Monday 3 August the *Victorious* left the Hudson River and set course for Portsmouth.

On the morning of Monday 10 August the *Victorious* returned to Spithead, and after clearing Customs she tied up alongside Pitch House Jetty some seven hours later. Joining the hundreds of wives and families who were ferried out to the ship at Spithead were 49 brides-to-be, whose prospective bridegrooms were all serving in the carrier. For Captain Coke the homecoming was tinged with sadness for, on the following day, he left the ship to take up an appointment at the Admiralty. The *Victorious*' new commanding officer was Captain H. R. B. Janvrin DSC RN, who had joined the Navy in 1929 and had qualified in 1938 as a Fleet Air Arm Observer. During the Second World War he had taken part in the attack on the Italian Fleet at Taranto and in the 1950s he had commanded HM Ships *Broadsword* and *Grenville*. Captain Coke left his old command in traditional manner - pulled in a car from the ship's side by Heads of Department, while the Royal Marines Band played 'Auld Lang Syne' and 'Will Ye No Come Back Again?' However, the return to Portsmouth and the change of command did not mean the end of the commission and on the morning of Tuesday 15 September the *Victorious* put to sea once again. After embarking her squadrons she steamed up the North Sea to take part in the exercises 'Blue Frost' and 'Barefoot' in the Norwegian Sea, just north of Trondheim and about 120 miles from the coast. The exercises were designed to test reaction to enemy landings in Norway, and they involved the destroyers *Armada*, *Dunkirk* and *Trafalgar*, together with the submarine *Talent* and RFA *Tideflow*. However, at 12.45pm on Tuesday 22 September, Skyraider 421 flown by Lt B. B. Hartwell and crewed by Sub-Lt H. D. Comber and Midshipman W. J. V. Walker, was reported as missing. The aircraft had last been heard from at 11.30am and so an air-sea search involving the destroyers and helicopters from the *Victorious* was initiated. In fact, the aircraft had run out of fuel and had ditched close to a fishing boat which was in the area. Fortunately, none of the crew were injured and they were picked up by the trawler which turned out to be Russian, before being transferred to its depot ship, *Atlantika*, the following morning. Meanwhile, the *Victorious* was co-ordinating an intensive search for the missing aircraft which involved not only the Royal Navy, but also the Norwegians, all to no avail for there was no sign of any survivors. However, news was received at the Admiralty, via Moscow, that all the crew members were safe and well and HMS *Urchin* was dispatched to bring them home. Later that month the First Sea Lord, Admiral Sir Charles Lambe, sent a letter of thanks to the Soviet Ambassador in London, expressing gratitude to the crew of the Russian trawler.

At the end of the exercise the *Victorious* returned to Portsmouth, where she arrived on the afternoon of Friday 2

October. Six days later the carrier's hangar was converted into the venue for Radio Luxembourg's version of 'Take Your Pick' with its presenter Michael Miles, who was also to make the game show more popular on the new Independent Television station. At 9.30am on Tuesday 13 October the *Victorious* left Portsmouth again, this time for flying exercises in Lyme Bay and the Channel before returning to Pitch House Jetty on Friday 23 October for a further seven days. When the carrier left Portsmouth at the end of the month she was bound for Gibraltar and the Mediterranean, and during flying exercises on Thursday 19 November a Scimitar ditched on take-off but, fortunately, the pilot was rescued by the planeguard helicopter. Next day the *Victorious* took part in exercises with the US Navy's 6th Fleet, including the USS *Saratoga*, and on Thursday 26 November she moored in Malta's Grand Harbour for a short stay of only four days. On Monday 30 November the *Victorious* left Grand Harbour for a very wet and cold visit to Marseilles, followed by a rough passage to Gibraltar and arrival at Portsmouth on Monday 14 December for an eagerly awaited Christmas leave, before the final deployment of the commission.

On the morning of 11 January 1960, as the *Victorious* was being manoeuvred by tugs from South Railway Jetty to D Lock for dry docking, she grounded with a slight list to port and remained stuck fast for seven hours until finally, at 10pm, she was refloated on the high tide and berthed safely in the lock. Soundings were taken of all the ship's double bottom tanks and fortunately no damage was found, and at 2pm on Monday 18 January she sailed from Portsmouth into the Channel. For the next six days the *Victorious* was to carry out deck landing trials of the new Blackburn NA 39, later to become the Buccaneer, and the aircraft made its first-ever deck landing at 12.28pm on Tuesday 19 January, when Lt-Cdr D. Whitehead AFC RN and Lt-Cdr E. A. Anson RN safely caught the wire and brought their aircraft to a halt in a textbook landing. Next day, at 8.17am, the NA 39 was launched from the carrier for the first time and just over 30 minutes later a second NA 39 flown by Commander Price and Lt-Cdr Evans made a safe landing aboard the *Victorious*. After spending Saturday 23 January at anchor at Spithead, the trials were wound up next day

The *Victorious'* first post-modernization commission ended on Friday 26 February 1960. Here, flying her paying-off pennant, she prepares to enter Portsmouth Harbour, whilst the outward bound *Queen Mary* steams past her.

(Fleet Air Arm Museum)

and the final tally for the Blackburn NA 39's performance was 31 successful launches and landings. That night the *Victorious* anchored at Spithead again, and at 2am on Monday 25 January there was an alarm when an oil tanker collided with a freighter five cables (1,000 yards) off the carrier's port bow. As the ship's company prepared to go to the assistance of the two vessels it was clear that both of them, locked together, were drifting close to her port side. The engine room was brought to immediate notice for steam and everyone held their breath as the two ships drifted down the port side, just three cables (600 yards) away. Later that day the *Victorious* went to sea for flying exercises, and after just 24 hours alongside Pitch House Jetty the carrier sailed for the Clyde, mooring off Greenock on the afternoon of Friday 29 January.

At 1.30pm on Wednesday 3 February the *Victorious* sailed for Icelandic waters where she took part in anti-submarine exercises with the frigates *Salisbury*, *Torquay*, *Paladin* and *Tenby*, in some very cold and rough weather conditions. At 4.32pm on Sunday 7 February, when the ship was in a position Lat 58°- 29'N/Long 02° - 08'W, a Scimitar piloted by Lt J. Westlake RN ditched shortly after take-off. Despite the fact that darkness had already fallen, a thorough search was made of the area, but unfortunately, with no result. Next day, off Lossiemouth, a memorial service was held for Lt Westlake. Five days later, at 3.40pm on Thursday 11 February the First Lord of the Admiralty spent two days on board to watch a flying display as ten Venoms, eight Scimitars and two Skyraiders were put through their paces, at one point in the middle of a severe snowstorm. The final foreign leg of the commission, a five-day stopover in Hamburg, began during the morning of Sunday 14 February when the *Victorious* entered the Elbe estuary. The visit had been well publicized and large crowds lined the banks as the carrier steamed slowly upriver from Cuxhaven. Throughout the stay hundreds of official visitors and thousands of the general public looked over the ship, and in turn the ship's company took the opportunity to make the most of Hamburg. It was said that the delights of the Reeperbahn competed favourably with the visit to the USA in the popularity stakes, and the Royal Marines Band Beat Retreat for the last time during the commission. The visit to Hamburg ended on Friday 19 February, and next day the ship passed the Dover Signal Station bound for the exercise areas in the Lyme Bay area. On the morning of Monday 22 February the Minister of Defence, Lord Carrington, and the First Sea Lord, Admiral Sir Charles Lambe, landed on board by helicopter to observe a flying display by Scimitars, and Sea Vixens from Yeovilton, before leaving for shore. Next day the *Victorious* moored in Plymouth Sound where the FOAC struck his flag, and on Thursday 25 February, with full ceremony, the carrier left Plymouth Sound for Portsmouth where she arrived alongside Pitch House Jetty at just after midday on Friday 26 February.

It was the end of a long commission and most of the ship's company took their farewells and left the ship which had been home for a very busy and successful two years. The *Victorious* herself was to have a short break before she recommissioned with a new ship's company.

Chapter Twelve

Crisis In The Persian Gulf

The end of March 1960 saw the *Victorious* high and dry in Portsmouth's D Lock and at this time work started on a rather unusual extension to the port side of the flight deck, forward of the port catapult. The odd-looking device was a bridle catcher and it consisted of a boom from which hung a nylon net, designed to scoop up the wire strop which connected the aircraft to the steam catapult. Before this invention the strops, costing £15 each, were lost in the sea after each launch, but the bridle catcher would allow them to be used four times. In an average month when about 1,000 aircraft were launched this amounted to a saving of approximately £11,250. On Friday 29 July the dry dock was flooded and the carrier was still in D Lock when the ship's company joined, and the ship commissioned on Tuesday 16 August. Although Captain Janvrin addressed the ship's company, it was decided to postpone the Commissioning Ceremony for a few weeks when it could be combined with another important event. On Thursday 18 August the *Victorious* was moved out of D Lock round to Middle Slip Jetty where a complex programme of storing ship and trials of the flight deck machinery was started. There was even a police investigation during the storing when large quantities of clothing were stolen from a storeroom. Finally, on Wednesday 14 September 1960 all work stopped for the day while the Commissioning Ceremony and 21st Anniversary celebrations of the ship's launching took place. The Lord Bishop of Portsmouth and Captain Janvrin conducted the service before the entire ship's company and their families as well as representatives of the squadrons and a number of Admirals, including Sir Manley Power, C-in-C Portsmouth, and Rear-Admiral Smeeton, FOAC. After the ceremony a birthday cake, weighing 250lb, was cut by Captain Janvrin who was ably assisted by the youngest member of the ship's company, Junior Seaman John Fletcher.

Thursday 22 September had been the date on which the refit was due to be completed, and Captain Janvrin had informed the ship's company that the *Victorious* would sail the following day. Everything was right on schedule and on Friday 23 September the *Victorious* left Portsmouth for steaming trials and DG ranging at Portland, during which

HMS *Victorious* in D Lock, Portsmouth Dockyard, in early 1960. Work has started to fit the bridle catcher forward of the port catapult. *(Fleet Air Arm Museum)*

every piece of machinery and equipment in the ship was tried and tested. Returning to Spithead on Saturday 24 September the carrier fuelled with AVGAS and embarked the 'boffins' and flyers of the Flying Trials Team for five days of flying trials in the Lyme Bay area. During that period aircraft from several squadrons, including 803, executed landing trials and the ship carried out the first of many replenishments from RFA *Tidereach*. While the carrier was alongside Pitch House Jetty, Portsmouth, between Saturday 1 and Sunday 9 October, she was fully ammunitioned and the flag of FOAC was hoisted for a week while he conducted the 'wash-up' of NATO's 'Exercise Swordthrust'. On Monday 10 October the *Victorious* left Portsmouth again to perform radar and radio trials, and for five days the ship went round and round in circles, but to good purpose for all the equipment was accepted and the session was concluded with full-power trials which achieved 29$\frac{1}{2}$ knots. There was a final long weekend in Portsmouth, during which time the squadron ground parties were embarked, and on the morning of Tuesday 18 October the *Victorious* sailed to embark the Scimitars of 803 Squadron, the Sea Vixens of 892 Squadron, the Gannets of 849 B Flight and six Whirlwind helicopters of 825 Squadron, before leaving for her work-up in the Mediterranean.

It had been arranged that the 1st Destroyer Squadron, which was also newly commissioned, would carry out anti-submarine exercises with the carrier on passage, but the weather in the Bay of Biscay turned nasty and they were cancelled. At one stage speed was reduced to eight knots in order to prevent damage to aircraft parked on deck, but blue skies in Gibraltar between Monday 24 and Wednesday 26 October put everything right with the first 'foreign' run ashore. After leaving Gibraltar and completing a day's flying off North Front, the *Victorious* made a fast passage to Malta for the first work-up period off the island. Following six consecutive days' flying the carrier entered Grand Harbour at just after midday on Friday 4 November, where the *Ark Royal* joined her a few hours later at an adjacent berth in Bighi Bay. Next day the flag of Rear-Admiral R. M. Smeeton, FOAC, was hoisted after he and his staff transferred from the *Ark Royal*. After ten days of self-maintenance the *Victorious* sailed on Tuesday 15 November for her second work-up period. The exercises took place to the south of Malta most of the time, but the ship did move down to the North African coast off Tripoli (Tarabulus) so that the squadrons could make use of firing ranges in Libya. It was whilst the ship was off Tripoli on Tuesday 22 November, that Gannet 425 ditched in the sea, with the tragic loss of its pilot, Sub-Lt R. Collins RN. Following the accident the remaining aircraft of 849 B Flight were grounded. Two days later the flight deck was crowded with goofers watching the lights as the *Victorious* steamed through the Strait of Messina en route for Naples, and next morning she anchored in the famous bay. In spite of a boat trip of well over a mile, large numbers of liberty men took advantage of sightseeing tours to Rome, Sorrento, Vesuvius and Pompeii. However, Naples Bay is very exposed from the west and strong winds on the last night stopped all boat traffic and stranded some 200 sailors ashore. Fortunately, most of them were able to get temporary accommodation on board USS *Cascade*, moored inside the breakwater, and they returned to the *Victorious* the next morning. The carrier sailed from Naples early on Tuesday 29 November for her third work-up period which consisted almost entirely of three major NATO exercises. The first of these, 'Exercise Royal Flush IV', took place in the western basin of the Mediterranean with *Hermes* and *Ark Royal*, and involved manoeuvres in close company on 3 December. Also taking part were the US Navy's aircraft carriers *Saratoga*, *Independence* and *Intrepid*. Following the exercise the three British carriers went their own ways on 4 December, with the *Hermes* leaving for the Far East to relieve the *Albion*, while the *Victorious* and the *Ark Royal* were to participate in an Army Support exercise code-named 'Pink Gin III' in the neighbourhood of Tobruk. This lasted for two days and then the *Albion* joined the two larger carriers for 'Exercise Decex' which took them westward past Malta and Cape Bon. It was during this exercise that a pilot lost control of a Scimitar as it left the catapult and he was forced to ditch over the port side. Fortunately, the pilot ejected successfully and he was picked up in double quick time by the SAR helicopter. The manoeuvres ended on the evening of Saturday 10 December, after which the *Victorious* set course for Gibraltar and a two-day break. However, the weather deteriorated rapidly and by dawn on 11 December the ship was hove-to in order to prevent serious damage to the aircraft which were parked forward. The wind remained at Force 8 all day, with some gusts reaching 51 knots, and boats and catwalks were damaged in the heavy seas. This in itself was bad enough, but at 2.26pm the rudder jammed at 5° to starboard. A ship without a rudder is an alarming state of affairs at any time, but at the height of a storm it presented a very dangerous situation and 'Not under Control' balls were hoisted. There was then a struggle to keep the ship head-to-wind in the gales using the main engines, but Captain Janvrin must have been relieved that it did not happen on the previous evening when the ship had passed within three miles of Cape Bon at 27 knots. Next day, during a lull in the storm, the ship's divers were able to make a preliminary inspection of the rudder and their reports made it clear that it would be impossible to free it without underwater cutting gear, and one of the *Albion's* Skyraiders was dispatched to Gibraltar to pick up the equipment. In the meantime, with weather conditions having improved, the *Victorious* got under way with a 3° list to starboard in order to assist the difficult task of

Alongside in Gibraltar in October 1960 following a rough passage from home waters.

(Fleet Air Arm Museum)

The *Victorious* and *Ark Royal* moored in Malta's Grand Harbour, November 1960.

(M. Cassar)

'Exercise Royal Flush IV' took place in the Mediterranean and it involved three British aircraft carriers. Here they make an impressive sight as *Victorious* leads the *Ark Royal* and *Hermes* at speed. *(Fleet Air Arm Museum)*

steering the ship by the main engines. That night, in gale force winds again, she made very slow progress towards Majorca and in the early hours of the morning, with a loss of vacuum in the port engine, the ship actually steered round in a complete circle. Finally, at 2pm on Tuesday 13 December the carrier anchored off Port Petro, on the south-east coast of Majorca, where divers could begin work. The *Albion's* Skyraider had already delivered oxy-hydrogen underwater cutting gear and with all the necessary diving gear and equipment having been made ready, the ship's divers were over the side within half an hour. Due to a combination of high speeds, vibration and heavy weather some fastenings had parted, allowing two large wooden blocks, covered with steel plate, to move and to jam in the few inches of clearance between the top of the rudder and the ship's bottom. It was therefore necessary to cut away the steel cover plates and remove the offending blocks. Working underwater against the swell and with vision restricted by their face masks, the divers faced an extremely difficult task, but finally, with the help of crowbars and their feet, they were able to dislodge the two wooden blocks and clear the obstruction, and at 4am on Wednesday 14 December the ship was able to weigh anchor and set course for Gibraltar. The ship's divers, who had been working for 15 hours at a time without a break, were then able to take a well-earned rest.

As the *Victorious* approached Gibraltar at 7am the next day, the First Sea Lord, Admiral Sir Caspar John, was embarked by helicopter and the carrier undertook an air defence exercise against the *Ark Royal*, before anchoring in Gibraltar Bay at noon. Four hours' leave were granted to each watch for last-minute Christmas shopping and at just before midnight the *Victorious* sailed for Portsmouth, bringing her back to the schedule which had been planned before the rudder jammed. Having flown off the aircraft to their shore bases, she anchored at Spithead during the evening of Sunday 18 December and at noon the next day she weighed anchor to steam up-harbour to berth alongside Middle Slip Jetty. Since recommissioning the *Victorious* had spent 62 days at sea and steamed 19,268 miles.

At 10.15 am on Wednesday 21 December 1960 Captain Janvrin was relieved by Captain J. M. D. Gray OBE RN, who had graduated from Dartmouth in 1931. He had served as a midshipman in the battleship *Nelson* and the cruiser *Enterprise* on the East Indies Station. When he was in the cruiser *Devonshire* the ship had rescued refugees from the Spanish Civil War. Having qualified as a specialist gunnery officer in 1938 he joined the small aircraft carrier *Hermes* and served in that ship until early 1942 when he was appointed gunnery officer of the cruiser *Spartan*, which was building at Barrow-in-Furness. His next ship, the cruiser *Orion*, had conveyed the Greek Government-in-Exile to Piraeus in October 1944 upon its restoration, and after the war he served in the battleship *Duke of York* and the cruisers *Glasgow* and *Swiftsure*. Promoted Captain in 1955, he commanded the frigate *Lynx* and this was followed by a tour in the Admiralty as Director of the Gunnery Division.

With all leave having expired on Thursday 5 January 1961, there was a hectic period of storing, ammunitioning and painting, interspersed with the wardroom's New Year's dance and a mammoth children's party in C hangar, before the ship sailed on Friday 20 January, having been delayed by an engine room defect. Next day the Vixens and Scimitars embarked in Lyme Bay and the ship steamed round to Cardigan Bay to carry out Firestreak missile firings. One Vixen pilot destroyed two targets in mid-air and then damaged his own aircraft and two parked Scimitars while landing on. At the end of this period the Gannets were embarked and the *Victorious* moored in Plymouth Sound to embark ground crews and to unload the damaged aircraft. The weather then delayed the ship's entrance into the Sound and so the Whirlwind helicopters were embarked off Portland, and after leaving Plymouth on Saturday 28 January she ran straight into heavy weather in the Bay of Biscay which resulted in a 24-hour delay in reaching Gibraltar. During the day spent in Gibraltar the flag of FOAC was hoisted and she sailed on the morning of Wednesday 1 February, in company with HMS *Blackpool* and RFA *Tidereach*, for the long, non-stop passage to Cape Town. The three ships remained together until the end of March and earned the nickname 'The Old *Vic* Touring Company'.

During the passage south flying exercises were carried out, followed by fast overnight passages to catch up with the *Tidereach*. At 10.42am on Wednesday 8 February, when the ship was just north of the equator, the last Vixen of a detail suffered a collapsed undercarriage on landing and it burst into flames. Fortunately, the crew were unhurt and they managed to jump out, but the aircraft was a write-off. As the ship steamed southwards the temperature increased and 'happy hour' was instituted when, between 4.15pm and 5.30pm, all flying stopped and the flight deck was declared open to all for a variety of sports and games ranging from deck hockey to judo. At about noon on Thursday 9 February the ship 'crossed the line' and that afternoon all work ceased as the Captain and all Heads of Department, together with several hundred novices, endured the time-honoured initiation ceremony. As the *Victorious* neared Cape Town she was met by HMS *Lynx* and SAS *Vrystaat*, and on Thursday 16 February she anchored in Table Bay close to the infamous Roben Island where a quick coat of paint was applied to the most weatherbeaten parts of the hull. Next morning, after weighing anchor, a number of prominent South African citizens and service personnel were embarked for a 'Shop Window' display. That afternoon, in spite of low cloud, many aspects of surface, underwater and aerial warfare were demonstrated. These were the days before the apartheid

Another view during 'Exercise Royal Flush IV', although the *Hermes* is almost obscured by smoke from the *Victorious* and *Ark Royal*. *(Fleet Air Arm Museum)*

policy, with all its brutality and injustice, was widely known about in Europe and South Africa was still buying arms from Britain. At sunset that evening the *Victorious* entered Cape Town to a very enthusiastic welcome, but she was also to become involved in unprecedented controversy in the UK brought about by the iniquities of apartheid. It soon became known to the British press that nine black naval personnel who had been serving in the carrier had been disembarked in Gibraltar and that some of them had been replaced by white ratings. This led to a storm in most British newspapers and the Admiralty defended its policy by stating that it would be wrong to, '...send an officer or rating to a country in which he would be liable to discriminatory treatment.' Despite the defence put forward by South African officials that a system of 'voluntary apartheid' had existed for three centuries and an assurance that black personnel serving with the US Navy had been graciously received and entertained, it was quite obvious that they would be subjected to the same indignities as the majority of the local population and they would be unable to enter a cinema or restaurant of their choice, or choose their compartment on a train. For many people in Britain the *Victorious*' visit to Cape Town highlighted one of the world's cruellest forms of discrimination.

However, most of the ship's company were unaware of the controversy at home and because of the overwhelming scale of the hospitality offered to the ship a special bureau was set up to cope with all the invitations from ashore. The ten days spent at Cape Town passed in a whirlwind of entertainment, parties in private homes and on board, barbecues and all kinds of sporting fixtures. The ship was open to the public on three days, and together with the private guests who visited the ship daily it was thought that about 25,000 people went on board during the ten days. On Sunday 19 February a new Queen's Colour was presented to the South American and South Atlantic Station by the British High Commissioner at a ceremony on the *Victorious*' flight deck. It was the first time that such a ceremony had taken place on board a ship, as normally a Colour is presented on sovereign territory, but in this case the expanse of the carrier's flight deck was the next best thing. The Royal Marines Band managed to avoid any political controversy by giving an excellent performance at a public park, which was enjoyed by people of all races.

At 9.15am on Tuesday 28 February the *Victorious* cast off her moorings and left Duncan Dock, Cape Town, with a course set for a far less popular port, Aden. Although the timetable did not allow the ship to stop at any other South African ports, the aircraft carried out fly-pasts at Hermanus, Port Elizabeth, East London and Durban, and at the first three the ship steamed close inshore. Whilst the *Victorious* was in the Mozambique Channel FOAC carried

Crash on deck. A Vixen catches fire after its undercarriage collapsed on landing. The pilot can be seen climbing out of his cockpit...

...Fortunately the crew were unhurt and firefighters soon doused the flames with foam. (*J. Bolton*)

This 'Crossing the Line' ceremony took place on 9 February 1961. Here 'King Neptune' and his 'court' await their victims. *(J. Bolton)*

out his Operational Readiness Inspection and HMS *Blackpool* was dispatched to Mombasa to collect mail and stores. Aden was reached on Saturday 11 March where HMS *Bulwark*, with 42 Commando embarked, had arrived from the Persian Gulf. After leaving Aden on Tuesday 14 March the *Victorious* rendezvoused with the *Hermes* on the next day and the two ships carried out a three-day exercise, 'Sea Sheikh', which included air defence and Army support in the Aden Protectorate. On conclusion of the exercise, during the afternoon of Friday 17 March, FOAC and his staff transferred to the *Hermes* by helicopter and she left for home while the *Victorious* set course eastward for Singapore, having assumed the operational role east of Suez. During the passage HMS *Blackpool* collected the mail from Colombo and the carrier refuelled from RFA *Resurgent* for the first time. On Tuesday 28 March the aircraft were launched to RAF Tengah in Singapore and next morning the *Victorious* berthed alongside the island's Naval Base, which she had last visited in December 1946. She had arrived in time for the Easter weekend having spent 27 out of the last 30 days at sea, and since recommissioning she had steamed 40,643 miles during 117 days at sea.

The first weekend at Singapore saw a large concentration of ships, including the *Bulwark*, HMAS *Melbourne* and a host of destroyers and frigates, all preparing for a big SEATO exercise, 'Pony Express', which was to be held later that month. While the *Victorious* carried out self-maintenance, the air squadrons, including those from the *Melbourne*, joined the RAF in the Singapore Air Show which was held at the civilian Paya Lebar Airport. On Wednesday 12 April the *Victorious* sailed to embark her squadrons then spent the weekend at anchor in Singapore Roads just south of the city itself. In the following week she joined up with the Far East Fleet for a work-up prior to 'Pony Express', and 825 Squadron operated from the *Bulwark*. On Wednesday 19 April Rear-Admiral M. Le Fanu, FO2 Far East Station, transferred with his staff from the *Belfast* and hoisted his flag in the *Victorious*. The work-up culminated on the weekend of Saturday 22 April in a well-organized banyan at Pulau Tioman where the *Victorious*' supply branch provided a jumbo barbecue. However, when the time came to weigh anchor on the evening of 23 April, four members of an expedition party were missing and HMS *Cassandra* was left behind to search for them. It turned out that their three-mile hike and 1,600ft hill climb had become something of a nightmare in the heat of a tropical rainforest, particularly as they had got lost.

During her passage north in the South China Sea the *Victorious* carried out fleet exercises and on Tuesday 25 April the Far East Fleet met up with the US Fleet from Subic Bay. The *Victorious* operated independently with an escort of two US destroyers, in support of an amphibious landing which was carrying out an assault on the beaches of North Borneo, with US Marines from the USS *Thetis Bay* and the Royal Marines from HMS *Bulwark*. The opposition consisted of submarines, RAF Canberras from Singapore and aircraft from the USS *Coral Sea*. The air defence phase was followed by ground support and at 4.35pm on Friday 28 April a Scimitar pilot, Lt G. C.

Edwardes RN, ejected after getting a fire warning. He was picked up by one of the *Bulwark's* helicopters and flown to the British Military Hospital at Singapore, but sadly he died of his injuries seven days later. The exercise ended on Wednesday 3 May with FO2 transferring his flag back to the *Belfast*, and next day most of the Far East Fleet steamed north for a Japanese cruise, while the *Victorious* returned to Singapore for an interim docking period. On 8 May 803 Squadron departed for RAF Butterworth, near Penang, 892 Squadron left for Tengah and 849 B Flight flew off to RAF Seletar on Singapore Island. The helicopters of 825 Squadron disembarked to RNAS Sembawang close to the Naval Base, and on Tuesday 9 May the ship secured alongside No 14 berth in the dockyard. Seven days later she was shifted into the King George VI dry dock, and the ship's company moved into every available billet in HMS *Terror* and the dockside accommodation. That same day came the very welcome news that the ship would return home for Christmas. As the fleet was away in Japan the dockyard was able to concentrate all its effort on the *Victorious* and a vast amount of work was carried out, including the repairing of the rudder which had behaved so badly five months earlier.

On Thursday 8 June the *Victorious* came out of dry dock and the ship's company moved back on board. Seven days later she sailed to embark the squadrons and to carry out flying exercises in the local area, followed by a weekend at Pulau Tioman and six days of flying and exercising. It was during this period, at 9pm on Tuesday 20 June, that Sea Vixen 216, which was carrying out night deck landing practice, stalled and crashed two cables off the port side of the ship in a position Lat 01° - 43'N/Long 105° - 07'E. The aircraft was being flown by Lt T. Gilbertson RN, with observer Sub-Lt R. T. Nelson RN, and despite an all-night search no trace of them was found. On Saturday 24 June the *Victorious* returned to Singapore where Rear-Admiral J. B. Frewen CB, the new FO2 Far East Station, hoisted his flag in the carrier. Two days later, in company with the RFAs *Tidereach*, *Reliant* and *Resurgent*, the destroyers *Cassandra* and *Carysfort* and the submarine *Teredo*, the *Victorious* left for Hong Kong, fitting in exercises en route. At 1.40pm that same day, at the spot where Sea Vixen 216 had crashed, the carrier stopped and wreaths were cast into the sea in memory of the lost aircrew. By the morning of Thursday 29 June the ships were almost halfway to their destination and flying, in rather inclement weather, had begun at just before 6am. However, at 7.30am an urgent signal was received from the C-in-C with the order to, 'Proceed to Persian Gulf with all dispatch.' Those who could pick up the BBC World Service were hearing reports of Iraq threatening to invade a tiny Arab state in the Persian Gulf called Kuwait.

Kuwait is a state of some 7,400 square miles in the north-western corner of the Persian Gulf with a population in 1960 of approximately 730,000. It is an undulating desert country bounded in the north and west by Iraq, in the south and south-west by Saudi Arabia and in the east by the Persian Gulf. British influence and interest started in 1899 when an agreement was signed with Kuwait, but with the discovery of oil in the 1930s Britain's interest intensified. Large-scale oil production in Kuwait began in 1946 and by 1960 it was one of the largest oil producing countries in the world. In 1961 it was agreed that the British would relinquish their protectorate and hand over power to the Sabah dynasty, who have ruled the country ever since. However, on 25 June 1961, just six days after the agreement came into force, General Abdul Kassim, President of Iraq, declared that Kuwait belonged to his country on the grounds that, under the old Ottoman Empire, it was a province of Basrah, which was indisputably Iraqi territory. It is clear that this claim was provoked by the widely held belief, particularly after Suez, that Britain lacked both the means and will to protect her interests abroad and to guard her allies. However, the British Government took Kassim's strongly worded statements seriously, as did the ruler of Kuwait, who requested military assistance from Britain. In the years leading up to 1961 there had been a succession of contingency plans for British military intervention in Kuwait, and as tension mounted in the Persian Gulf these were put into action and troops were soon on their way to the area. As it happened the commando carrier HMS *Bulwark* was already in the Arabian Sea and with 600 men of 42 Commando Royal Marines on board she hastened to the Kuwaiti coast.

As soon as she received the signal the *Victorious* set course for Singapore at 26 knots, whilst continuing to carry out flying exercises. Next day, as 2-inch rocket projectiles were being loaded onto a Sea Vixen, one of the missiles exploded, badly injuring CPO Carter and NA Ellis. They were quickly ferried by the *Carysfort* to Singapore where they could be taken to hospital, and that evening the *Victorious* steamed north through the Strait of Malacca. Early on 2 July she refuelled from RFA *Wave Ruler*, which was waiting off the coast of Sumatra, and course was set at 23 knots across the Bay of Bengal. Very little flying was done during the following few days as the south-west monsoon whipped up some very heavy seas and high winds, but on Saturday 8 July the *Victorious* reached the shelter of the Gulf of Oman where she refuelled from RFA *Orangeleaf*. Later that day she entered the stifling heat of the Persian Gulf where it was over 110°F in the shade, but the problem on the flight deck was that there was very little shade, while down in the machinery spaces temperatures of well over 125°F were recorded. There was little relief to be found in the sea where it registered 92°F and the air was full of sand blown about by the strong winds. It was a very inhospitable climate, particularly in those days before air-

Crowds gather to watch the *Victorious* enter Cape Town Harbour during the early evening of Friday 17 February 1961.
(*J. Bolton*)

Four very popular visitors during an open day at Cape Town.
(*J. Bolton*)

A Sea Venom from HMAS *Melbourne* lands on board the *Victorious* during 'Exercise Pony Express'.

(*J. Bolton*)

A Sea Vixen carries out touch-and-go circuits.

(*J. Bolton*)

conditioning. By this time there were 6,000 British troops ashore, the *Bulwark* and her commandos were standing by, and the *Victorious* had taken over the air defence of the tiny state. The ship's routine developed into a steady cycle of three or four days' flying, with a day for refuelling and one for maintenance. Flying would start early in the morning and it would continue until 1pm when all work, apart from essential watchkeeping, would cease until the evening when it could be resumed. This system did help to relieve the worst effects of the terrific heat, but nevertheless conditions on board were very trying. Most of the ship's company suffered from prickly heat - that maddening, itchy inflammation of the skin. Fortunately, there were very few cases of heat exhaustion, even though the cooks were able to fry eggs on the flight deck. On Monday 10 July there was an accident involving Scimitar 151 when its brakes failed and it ditched over the port side. Fortunately, the pilot, Lt D. S. McIntyre, was recovered unhurt. At dawn on Friday 14 July, which was the day thought most likely for an Iraqi invasion, the ship went to 'Action Stations', but as there was no sign of movement ashore it turned out to be a day of routine flying.

At 12.42pm on Monday 31 July the *Victorious* rendezvoused with HMS *Centaur* which had steamed out from the Mediterranean, having cancelled a trip to the USA, to relieve her in the Persain Gulf. After Admiral Frewen had transferred his flag to the *Centaur* the *Victorious* set sail for Mombasa at 23 knots, and with the south-west monsoon still blowing hard, once she rounded the eastern tip of Arabia and left the shelter of the Gulf Of Oman, the air and sea temperatures dropped by 12 degrees in just 30 minutes. The relief throughout the ship was tremendous. With no flying en route the ship was able to keep up a steady speed and she entered Kilindini Harbour, Mombasa, on the morning of Tuesday 8 August. Since leaving Singapore on 15 June the ship had steamed some 20,877 nautical miles. In the event the Iraqi leader did not invade Kuwait and it was to be almost 30 years before an even more ruthless leader of that country decided to plunge his country into a disastrous war by doing so.

The Kuwait crisis had played havoc with the Navy's carrier programme in that the *Centaur* would be very late home for docking and leave, which in turn would mean a delay in the *Victorious* being relieved by her. It was also clear that the *Victorious* would have to remain on the Middle East Station as long as there was a requirement for an aircraft carrier to be within six days' steaming of Kuwait. However, in the meantime, after the Persian Gulf, Mombasa was a real haven and 48 hours' leave was granted to each watch, with safari expeditions into the game reserves being the most popular pursuits. Back on board meanwhile, the ship was washed and painted from the truck to the waterline, and all the aircraft were thoroughly overhauled. On the morning of Monday 21 August, with full ceremony, the *Victorious* left Mombasa for flying exercises off Malindi and it was expected that she would return for the weekend. However, during the week there was civil unrest in Zanzibar and two of 825 Squadron's helicopters were sent to support the Police, with the jets flying low over the offending villages. The Whirlwinds would have stayed for longer but, quite unexpectedly, the ship was released from her obligations in the Middle East and she was ordered back to Singapore in order that the dockyard could undertake work on the flight deck. On the way she collected two Sea Vixens which had been left in Aden by the *Centaur*, and on Friday 15 September she arrived back in the Naval Base at Singapore. Three weeks were spent alongside No 8 berth there and these were devoted to clearing the backlog of maintenance which had built up. It was Thursday 5 October when the ship was ready for sea again, to take part in a series of exercises called 'Fotex', which involved a selection of ships from the Far East Fleet and Commonwealth units, including the destroyers *Caesar*, *Cavalier* and *Caprice*, and the Australian destroyers *Quiberon* and *Vampire*. The *Victorious* had only a minor role in the exercises, with the helicopters joining in the anti-submarine phase and the jets simulating attacks on the fleet. On Friday 13 October the carrier left the Singapore area for Hong Kong, and this time she actually reached the colony although, as she was far too big to go alongside the dockyard wall, she was secured to No 1 buoy in the middle of the narrowest part of the harbour. Fortunately, an excellent ship-to-shore ferry service overcame most of the inconvenience of not being alongside and, despite the fact that it was the typhoon season which restricted leave somewhat, the weather stayed fine. The visit came to an end on Wednesday 25 October when, in company with the *Cavalier*, the *Victorious* sailed for the Philippines to join the USS *Ticonderoga* in 'Exercise Crosstie'. Next day the two carriers flew strikes against each other, and at 1.57pm on Friday 27 October Sea Vixen XJ 519, which was about to be launched, broke her holdback bridle and trundled over the bows into the sea. Fortunately, the crew were rescued, bruised but otherwise unhurt, by the SAR helicopter. On the morning of Saturday 28 October all the ships tied up at the US Naval Base in Subic Bay for the weekend, but on Monday they were at sea once again and the two carriers cross-operated their aircraft. On Tuesday 31 October, when a Sea Vixen was landing on the *Ticonderoga*, its nose oleo collapsed, but again the crew were not hurt. The exercise ended that evening and, after transferring the damaged Vixen by lighter, the *Victorious* set course for Singapore where she arrived on the morning of Saturday 4 November.

After a brief stay in Singapore, to complete the Christmas shopping, the *Victorious* left the Naval Base for Aden on Tuesday 14 November, in company with HMS *Chichester*, which was to act as planeguard for two days.

However, early on Thursday 16 November, just when the carrier had passed the Nicobar Islands, she was diverted to Mombasa so that 825 Squadron's helicopters could assist with flood relief work at Lamu Island, 180 miles north of the city of Mombasa. The area had been struck by heavy rains and the island had been cut off from the mainland, but there was an audible sigh of relief on board when it was announced that the ship's return to the UK would not be delayed. The *Victorious* arrived alongside 9 & 10 berths of Kilindini Harbour on Wednesday 22 November and personnel and equipment were disembarked. Two helicopters flew ashore and all the men and supplies arrived at Lamu the next day, as the carrier herself left Mombasa for a short 12-hour stay at anchor in Aden Bay, before setting course for Suez. It was just before 8pm on Friday 1 December when the *Victorious* met the *Centaur* in the Gulf of Suez and collected two helicopters of 824 Squadron in place of those from 825 Squadron which the *Centaur* would take over in Mombasa. Before they parted company FOAC and his staff transferred to the *Victorious* and his flag was rehoisted. Two hours later she anchored off Newport Rock to await a northbound convoy through the Suez Canal. There had been some concern as to whether the ship, with her fully angled flight deck, would be able to navigate the canal, but as the *Hermes* had made it safely in December 1960 it was decided to attempt it. She entered the canal at first light on Saturday 2 December, leading the convoy, and it was soon found that the ship handled well and the transit was completed successfully at 6.30pm that evening. There had been no problems and she even passed the El Ferdan swing bridge, which carries an important railway line, with plenty of room to spare. For the entertainment of the ship's company four 'Gully-Gully' men had been embarked at Suez and they performed extremely clever tricks with their tiny yellow chicks.

Tuesday 5 December saw the *Victorious* approaching Malta and for three days she took part in a NATO exercise code-named 'Royal Flush V' with the destroyers *Battleaxe* and *Crossbow*, the cruiser *Tiger* and units of the US Sixth Fleet. On Sunday 10 December she entered Grand Harbour and two days were taken up with transferring stores and equipment to the *Ark Royal* while the latter's Whirlwind helicopters of 815 Squadron were embarked in the *Victorious* for the passage home, as well as two damaged Gannets from Hal Far. Finally, at 9am on Tuesday 12 December the *Victorious* left Malta for a fast passage to Portsmouth and two days later she passed Europa Point. She anchored in Spithead at just before midnight on

The *Victorious* in the King George VI Dry Dock at Singapore Naval Base in May 1961. The ship's company had moved into shore accommodation.

(J. Bolton)

Leaving Singapore Naval Base with full ceremony, 1961. *(J. Bolton)*

At speed in the Persian Gulf, July 1961. *(Maritime Photo Library)*

A Scimitar is lost over the port side in the Persian Gulf after its brakes failed. Fortunately, the pilot was rescued safely. (*J. Bolton*)

Servicing a Sea Vixen in the searing heat of the Persian Gulf. (*J. Bolton*)

Following the Kuwait crisis of 1961 the *Victorious* steamed south to the more welcome haven of Mombasa. Here she enters Kilindini Harbour.
(Fleet Air Arm Museum)

A Scimitar about to be launched.
(J. Bolton)

Sunday 17 December and all next day was spent clearing Customs but, at last, on the morning of Tuesday 19 December she weighed anchor and, through the lifting mist and under a weak December sun, she steamed up-harbour. Waiting for her at South Railway Jetty were hundreds of parents, wives, children and sweethearts, and after she secured alongside there were affectionate reunions as they streamed on board, with refreshments being served in one of the hangars. She had been away from Portsmouth for 333 days and during that time she had steamed 82,000 nautical miles at an average speed of 17$^1/_2$ knots, but now it was time for some Christmas leave before starting the final lap of the commission.

The first leave period ended in mid-January 1962, and a very successful children's party was held in the hangar where 215 youngsters were entertained in the traditional manner, and on 16 January tugs towed the *Victorious* to Middle Slip Jetty so that some use could be made of the large dockside crane. On Monday 22 January everyone was reminded of the ship's real function when the Whirlwinds of 825 Squadron started flying again, and on the same day Rear-Admiral R. M. Smeeton was succeeded as FOAC by Rear-Admiral F. H. E. Hopkins, the latter hoisting his flag in the *Victorious* on the following day. On Saturday 27 January a Families Day was held on board and on Monday 5 February, with full ceremony, the carrier left Portsmouth to embark the squadrons in Lyme Bay and to carry out exercises in the Channel. During the afternoon of Tuesday 13 February a Whirlwind helicopter ditched in Lyme Bay, but three of the crew were rescued by another helicopter, and the seaboat rescued the fourth. Two days later came a four-day visit to Brest, followed by more exercises in the Channel and a 12-day self-maintenance period in Gibraltar, which ended on Tuesday 13 March. After leaving the colony the *Victorious* carried out a series of trials with the 1st Destroyer Squadron before arriving at Vigo in north-west Spain on Saturday 17 March for four days alongside the Transatlantic Passenger Terminal. Leaving Vigo on the afternoon of Wednesday 21 March she took part in NATO's 'Exercise Dawn Breeze VII' in the Atlantic, during which she carried out cross-decking operations with the French aircraft carrier *Clemenceau* and played host to the First Sea Lord and the C-in-C of the French Navy. The exercise ended on Wednesday 28 March and two days later the aircraft flew off to their parent stations and the *Victorious* set course for Portsmouth. After a night at anchor at Spithead she secured alongside South Railway Jetty, Portsmouth, at just before midday on Monday 2 April 1962. It was the end of another successful commission, during the 594 days of which she had steamed 114,000 nautical miles, and had spent 323 days at sea.

Alongside and dressed overall at Subic Bay. *(Fleet Air Arm Museum)*

The *Victorious* arrives at Portsmouth on 2 April 1962 to pay off at the end of another commission. *(Fleet Air Arm Museum)*

Chapter Thirteen

Two Commissions In The Far East

At the end of April 1962 the *Victorious* was moved into No 3 Basin at Portsmouth Dockyard, and on 1 May Captain Gray left the ship as she was taken over by the dockyard for a 14-month refit. During this time some visible alterations were made to the ship's appearance, notably the removal of the bridle catcher and the fitting of a new wind recorder on the starboard bow and new landing sights on the flight deck. The flight deck itself was strengthened to cater for the Blackburn Buccaneer strike aircraft, while internally the most important improvement was the extensive installation of air-conditioning which made life on board much more comfortable, particularly in the mess decks.

On 19 November 1962 Captain P. M. Compston RN took command and a week later the whole country was plunged into the most bitterly cold winter for over 80 years, ushered in by the heaviest snowfalls of the century which froze solid for four months. The temperatures were so low that the sea water in B Lock completely froze over, and the flight deck, which was covered in several inches of solid ice, was like a skating rink. With the hangar lifts down, the huge compartment became an Arctic wind tunnel and with no heating in the ship the firemain system continually froze up. For the duty watches who had to live on board it was almost like living in a giant refrigerator. One officer recalls sitting in the Admiral's dining cabin one duty night watching television whilst wearing a heavy greatcoat and thick scarf, and still shivering with cold.

The ship's new commanding officer, Captain Compston, had joined the Royal Navy in 1937 and he had quickly specialized in flying duties. During the Second World War he had served in the ill-fated aircraft carrier *Ark Royal*, the battleship *Anson* and another aircraft carrier, HMS *Vengeance*. After the war he served in the light fleet carrier HMS *Warrior*, which was on loan to the Canadian Navy at the time, and after being promoted Captain in 1955 he took command of the frigate HMS *Orwell*. After a posting to the British Embassy in Paris, where he was the Naval Attaché, he was appointed to command the *Victorious*.

Most of the ship's company joined in early June 1963 and the refit ended on 12 June, but the Commissioning Ceremony was not held until 17 days later on Saturday 29 June in C hangar. The guest of honour at the ceremony was Vice-Admiral Sir John Hamilton, the Flag Officer Air (Home), and the FOAC, Rear-Admiral D. C. E. F. Gibson, flew 6,000 miles from HMS *Ark Royal* in the Far East in order to attend the ceremony. However, bad weather on the last few miles forced his helicopter to return to Yeovilton, and he had to complete the journey to Portsmouth by car, missing the ceremony by about an hour. He was, however, able to take the opportunity to look round the ship. At 9am on Tuesday 2 July, with full ceremony, the *Victorious* left Portsmouth to carry out trials of the ship's machinery in the Channel. These soon brought to light engine room

On 29 July 1963 the *Victorious* left Portsmouth following her refit, to carry out flying trials.
(Wright & Logan)

A pilot's eye view of the flight deck on landing. *(Fleet Air Arm Museum)*

In this view of HMS *Victorious* a Supermarine Scimitar is about to be catapulted off. *(Fleet Air Arm Museum)*

Manoeuvring at speed, and making a lot of black smoke. (J. Harper)

defects, for on the following day her departure from Weymouth Bay was delayed by four and a half hours. A few days later the full-power trials were disrupted by fog, but on Thursday 11 July she returned to Portsmouth for foreign service leave, and for the dockyard to rectify the faults which had shown up. Just over two weeks later, on Monday 29 July, following a visit from the C-in-C Portsmouth, the *Victorious* left Portsmouth once again for flying trials, and during the next 11 days there were long sessions of deck landing practice for the Buccaneers of 801 Squadron, the Vixens of 893 Squadron and Gannets from the Operational Flying School. During night flying operations many visitors came to look at the new red floodlighting on the flight deck, and at 7.50am on Friday 2 August there was a close call when a Vixen was being launched, for just as it reached the end of the catapult the observer ejected from the aircraft because he thought it had suffered a wing-fold failure. He landed close alongside the ship, the engines were put full astern and the seaboat was lowered, and, fortunately, he was rescued safely. A few days later a Buccaneer landed short of the wires, bounced over them and only just managed to avoid ditching as he went round again. Then on Thursday 8 August came a signal from the Admiralty requiring the ship to sail for the Far East seven days earlier than scheduled. Having completed her home leg of the General Service Commission, it had been planned that she would spend 12 months east of Suez so, although the signal took everyone by surprise, it was not a major alteration to her programme. The departure had been brought forward to maintain the Navy's carrier strength east of Suez while the *Ark Royal* was undergoing repairs to faulty engine machinery which had caused her to withdraw from 'Exercise Fotex 63'. The *Ark Royal's* breakdown also led to a change in the sailing programme of HMS *Hermes* which had been east of Suez since November 1962, and she was ordered not to steam through the Suez Canal until relieved by the *Victorious*.

On receiving the signal the *Victorious* returned to

...A Vixen on the catapult...

...and about to become airborne.
(J. Harper)

A Buccaneer on the catapult.

(J. Harper)

Portsmouth and there then followed four hectic days during which squadron personnel were rounded up from all over the country, mechanical defects were rectified and stores were rushed on board. With farewell parties having been abandoned, the carrier left Portsmouth on the morning of Wednesday 14 August, and as she steamed down Channel the Buccaneers and Vixens of 801 and 893 Squadrons were embarked, together with the Wessex helicopters of 814 Squadron and the Gannets of 849 A Flight. Hardly had the ship's company got over the surprise of the early departure than Gibraltar was being left astern, and white knees were much in evidence as tropical rig appeared. As the *Victorious* steamed through the Mediterranean Sea there were brief glimpses of the North African coast and occasionally an island, but the first pause was off Malta on the afternoon of Sunday 18 August when 'hands to bathe' was piped. However, one member of the ship's company injured himself diving from the ship, and had to be landed at Bighi Hospital. Meanwhile, the *Victorious* continued her voyage to Port Said where she arrived on the afternoon of Tuesday 20 August and dropped anchor. It was in the early hours of the next morning that she weighed anchor and started her transit of the Suez Canal, and by 5pm that afternoon she had cleared Port Suez, where everyone on board the *Hermes* was eagerly awaiting her arrival. In the Red Sea, as the temperature rose steadily, the flight deck became a popular sunbathing area and those who had served on the previous commission appreciated the recently installed air-conditioning. On the morning of Saturday 24 August the ship arrived off Aden and it was back to business on the flight deck, when two Gannets were launched to RAF Khormaksar. Next day there was a full day's flying, the first since the ship had left Portsmouth, and on Monday 26 August she moored in Aden Harbour to undertake a nine-day self-maintenance period. During the stay in this inhospitable colony, 60 schoolchildren visited the ship and the King's Own Scottish Borderers came on board to Beat Retreat on the flight deck. On Monday 2 September the ship was opened to visitors for the first time in the commission, and some 900 people came out by boat from Steamer Point to look round the carrier.

After leaving Aden on the morning of Wednesday 4 September, and refuelling from the *Tidereach*, the ship's work-up began in exercise areas off the colony. During flying operations on Tuesday 10 September there was a visit from a Parliamentary delegation who toured the ship. That same evening, at 7.30pm, contact was lost with one of 814 Squadron's helicopters and a few minutes later it was confirmed that it had ditched in the sea. The planeguard

A very smart *Victorious* at the Singapore Naval Base. (*J. Harper*)

frigate HMS *Eskimo* was soon on the scene and she rescued the three members of the crew, two of whom were returned to the *Victorious* while the third was taken to hospital at RAF Khormaksar in Aden. On Friday 13 September the ship returned to Aden for a brief three-day visit before setting course for Singapore and the Far East Fleet. At 11.30am on Tuesday 24 September she rendezvoused with the *Ark Royal* and carried out exercises with her during the morning, after which the two ships parted company and went their separate ways - the *Ark* to the Persian Gulf and the *Victorious* to Singapore, where she arrived the following morning.

In December 1962 an armed revolt had broken out in the Sultanate of Brunei, a small oil-rich country west of the colony of North Borneo, and although this had been dealt with quickly and efficiently by a force of Royal Marines and Gurkhas, with the help of the commando carrier HMS *Albion*, it was clear that there going to be serious problems with Indonesia and her president, Achmed Sukarno. It was strongly suspected that he had been behind the Brunei rebellion, as he was vehemently opposed to the Federation of Malaysia which was being negotiated to bring together Sabah, Sarawak and Malaya. The Federation was not a popular idea in Brunei, and it was here that Sukarno saw an opportunity to foment trouble. Although the rebellion was a failure, in April 1963 groups of Indonesian 'volunteers' began to infiltrate across the border into Sabah and Sarawak, triggering one of Britain's colonial 'bushfire' wars which became known as the 'Confrontation' with Indonesia, and which lasted until August 1966. The first real escalation of the crisis came on 28 September 1963, when about 200 Indonesian raiders crossed the border and attacked a small outpost at Long Jawai which led to incidents all along the border. Although the Army, with the dedicated support of HMS *Albion*, was able to cope with the aggression, it was not known how much further the Indonesians would go and it was expedient therefore for the *Victorious* to remain in the vicinity of Singapore for the foreseeable future. On Saturday 12 October the First Lord of the Admiralty, Lord Carrington, and the C-in-C Far East Station, Admiral Sir Desmond Dreyer, visited both the ship and the squadrons at RAF Tengah in the north-west corner of the island. Four days later, on Wednesday 16

The *Victorious* enters Subic Bay in November 1963.
(*J. Harper*)

October, the *Victorious* left Singapore to resume her work-up and to carry out engine trials in company with the C-in-C's dispatch vessel *Alert*, the frigate *Loch Alvie*, the destroyer HMAS *Quiberon* and the submarine *Amphion*. All went well for three days, but early on 19 October the failure of the auxiliary feed pump on the centre shaft unit meant that the carrier would have to go into a dockyard for repairs. A deputation of senior officers assembled to discuss the problem, and to the delight of the ship's company they decided to send the *Victorious* to Hong Kong. On the morning of Wednesday 23 October the squadrons disembarked to Kai Tak airfield, and the *Victorious* followed close behind to No 1 buoy which was strategically placed in the harbour between Wanchai and Kowloon. While the dockyard carried out the repair work, Jenny and her Side Party gave the ship a new coat of paint, a party was held for orphaned children and a good deal of money was spent in the local shops. At the end of the stay there were many sad faces when the ship left Hong Kong for Subic Bay to carry out joint exercises with the US Navy.

Throughout this period President Sukarno of Indonesia continued his belligerent propaganda campaign against Malaysia, although it was accompanied by only minor acts of aggression. It was obvious, however, that the presence of the *Victorious* was a very effective deterrent, for on one occasion when the ship was in the South China Sea it seems that President Sukarno's arrival by air at Manila was delayed for two hours because his aircraft was diverted to keep it out of range of the carrier's fighters. During the exercises with the US Fleet there was a two-day break when the *Victorious* went alongside Alava Wharf in the US base at Subic Bay, and the carrier was off Manila when, at 3.40am on Saturday 23 November news was received of the assassination of President Kennedy. Three days later the *Victorious* was back alongside the Naval Base at Singapore for a short visit before, on Friday 29 November, she sailed for 'Exercise Kit Kat', a joint exercise testing the air defence for the island of Singapore which was carried out off the east coast of Malaya with a great deal of publicity for President Sukarno's benefit. At 10.50pm on Thursday 5 December, during night flying exercises, a Sea Vixen crashed on landing but, fortunately, the crew were unhurt although their Firestreak missile was lost overboard. Five days later the *Victorious* returned to Singapore for a period of self-maintenance and for the Christmas festivities.

It was during her period alongside that a major change in the ship's programme was announced. Because of the Indonesian 'Confrontation' it was decided that the *Victorious* would not return home to Portsmouth in August 1964 as had been planned, but instead her refit would be carried out in Singapore and she would also recommission there, with the majority of the ship's company being flown

With the ship's company manning the flight deck the *Victorious* enters Singapore Naval Base.
(*J. Harper*)

The *Victorious* and the *Hampshire* refuel from RFA *Tidespring*. (*J. Harper*)

HMS *Victorious*, the famous Star Ferry and a local Chinese junk give this photograph in Hong Kong Harbour plenty of atmosphere. (*Fleet Air Arm Museum*)

The carrier moored at No 1 buoy in Hong Kong Harbour. She was moored between Wanchai and Kowloon with the Peak providing a backdrop to the photograph.
(Hong Kong Government Information Service)

Helicopter operations.
(J. Harper)

home over a period of three months and their reliefs joining the ship the same way. For some this was rather a blow for they had bought bulky presents such as camphorwood chests, which they would be unable to take home by air, and there were others who did not want to fly - after all they were in the Navy - and requests to stay on for the next commission were submitted. On Christmas Eve a carol service was held on the flight deck for the fleet and the dockyard, and some 2,000 attended to sing. Many members of the ship's company were spending their first Christmas away from home, and despite the unseasonal weather it was celebrated in the traditional manner. The *Victorious* sailed from Singapore on Thursday 2 January 1964 to carry out flying exercises off the east coast of Malaya, which included a joint army support exercise 'Cocktail' in company with HMNZS *Taranaki* and HM Ships *Diana*, *Lincoln* and *Andrew*. However, world events were about to change the course of the *Victorious*' schedule once again.

On 9 December 1961 the East African colony of Tanganyika was granted independence under Julius Nyerere, who wished to combine socialist ideals with the traditions of African village life. Just off the coast of Tanganyika lay the island of Zanzibar which had undergone constitutional reform during the late 1950s and political parties had formed more or less along racial lines with the Zanzibar National Party being the dominant force under Arab leadership. The African majority had much less influence and shortly before Independence Day on 10 December 1963 the ZNP lost the vital support of the trade union movement. Soon after independence the Sultan of Zanzibar was overthrown by the underprivileged African majority, who also massacred their Arab co-nationals. These bloody events in Zanzibar were as worrying to President Nyerere as they were to the British Government who were the ex-colonial power, and he not only granted the deposed Sultan asylum in Tanganyika, but he also opened negotiations with the rebel Government to try to effect a union between his country and Zanzibar. Two Royal Navy warships were sent to Zanzibar in order to help with the evacuation of British nationals. Initially it seemed that the disturbances in Zanzibar were internal racial disputes, but the unrest throughout the newly independent African states spread to Tanganyika itself, and also to Kenya and Uganda where it manifested itself in the form of mutinies in their respective armies. Following independence the armies of both countries remained very much as they had been under the colonial governments, with the private soldiers and NCOs being led by British officers. Despite the fact that measures were in hand to train capable local men for positions in command, the process was not fast enough to satisfy the aspirations of some of the troops and on 20 January 1964 men of the 1st Battalion Tanganyikan Rifles mutinied against their British officers at Colito Barracks, Dar es Salaam. President Nyerere immediately requested British military assistance to quell the unrest and at Aden 600 men of 45 Commando, Royal Marines, embarked in the *Centaur* which had arrived from the UK. In the meantime, the *Victorious* was on passage to Mombasa to carry out her Operational Readiness Inspection in the area and on 19 January 1964 she held her 'Crossing the Line' ceremony, arriving off Gan the next day for flying exercises. Then on Thursday 23 January she was ordered to sail for Mombasa 'with all dispatch', and although her role in the East African mutinies was uncertain at that time, during the passage landing parties were trained for a landing by helicopter. She arrived off Kilindini Harbour on Sunday 26 January and that night she steamed close inshore with her island superstructure and ensign staff floodlit. Next day she entered the harbour and on Tuesday 28 January she was ordered to Dar es Salaam to take over from HMS *Centaur* which had landed 45 Commando three days earlier. The *Victorious* arrived off Dar es Salaam on the morning of Wednesday 29 January, and over the next few days she embarked a very buoyant 45 Commando, whose efficient handling of the mutiny had made them very popular with all the communities in the city. Two RAF Belvedere helicopters landed on board and conditions were very cramped as the Marines bedded down in A and B hangars. Although there was no leave granted there were banyan parties to nearby Makatumbe Island which were very popular and the Royal Marines Band went ashore on a very successful public relations duty. Members of the press were entertained on board and Captain Compston also received officials from the Tanganyikan Government. Apart from launching two Gannets as a Communications Flight there was no fixed-wing flying as literally every available square inch was taken up by Commandos parading, training or sleeping. Lifts full of Marines in battle order would ascend from the hangars and they would rehearse hopping in and out of helicopters with all their gear, or they would practise small arms shooting, shattering targets in the sea. Finally, on Friday 7 February, with the tensions ashore having eased sufficiently, the *Victorious* weighed anchor and set course for Mombasa where she arrived the next day. Before her arrival, most of the fixed-wing aircraft were launched to Nairobi Airport, for the ship's three weeks at Kilindini were to be spent carrying out essential maintenance tasks. During the afternoon of Sunday 9 February, HMS *Albion* arrived at Mombasa and 45 Commando began the task of transferring themselves and their equipment to the commando carrier, together with the two Belvedere helicopters, in readiness for their return to their base in Aden.

After leaving Mombasa on Saturday 22 February and embarking her squadrons, the *Victorious* set course for the area between the Nicobar Islands and Penang, where she was to take part in 'Exercise Jet 64'. During the passage

A Gannet is prepared for launching from the starboard catapult.
(J. Harper)

A Buccaneer lands on.
(J. Harper)

HMS *Victorious* and HMS *Albion* at Kilindini Harbour, Mombasa, following the army mutinies at Dar es Salaam. HMS *Salisbury* is in the background.
(N. McCart)

there was no time for flying except by the helicopters which had to get their anti-submarine equipment serviceable once again. However, as she approached Langkowi Island the Vixens made a long-range sortie over the anchorage where the rest of the 'Jet' fleet were anchored. After arriving at Langkowi on the afternoon of Sunday 1 March FO2 Far East Station was embarked and next day the exercises, which involved the Indian aircraft carrier *Vikrant* (ex-HMS *Hercules*), as well as Australian, New Zealand, Canadian and Royal Navy ships, got under way. The exercises kept the fleet busy for over two weeks, with only a weekend break at Langkowi, and when they were concluded on Thursday 19 March all the participants retired to the Naval Base at Singapore for the inevitable 'wash-up' and a self-maintenance period for the *Victorious* which took her through to early April. Sadly, the stay was overshadowed by the death of Sub-Lt (S) G. Ellis in a road accident in which his colleague Sub-Lt A. J. Stewart was seriously injured. On Friday 3 April the *Victorious* left Singapore to embark her squadrons and to carry out deck landing practice before returning to the base on 9 April for four days. On the morning of 13 April she sailed again in company with the Australian ships *Parramatta* and *Yarra*, and HM Ships *Hampshire* and *Lincoln*, all in close attendance. The *Hampshire* was the first of the new class of destroyers to reach the Far East, and after some private flying the ships rendezvoused with HMS *Centaur* to take part in another air defence exercise designed to test the Singapore defences. During the operations the FOAC, Rear-Admiral H. R. B. Janvrin, who knew the *Victorious* as well as anyone else on board, visited for a day from *Centaur* as he was keen to see

all the changes which had been made during her last refit. On Sunday 19 April the *Victorious* sailed for Hong Kong once again and on arrival two days later she laid on a 'Shop Window' display for local VIPs, on completion of which the aircraft left for Kai Tak and the carrier secured to No 1 buoy. The visit coincided with Anzac Day and men from Australian and New Zealand ships in Hong Kong harbour joined a large parade at the war memorial in Victoria, which was concluded by a fly-past of RAF Hawker Hunters and the squadrons' Vixens and Buccaneers. They were followed by a lone Wessex of 814 squadron, which was crewed by both an Australian and a New Zealander, streaming two Anzac flags. Once again *Victorious* received a new coat of paint, courtesy of Jenny's Side Party, and when the carrier departed for Japan on Wednesday 6 May, with full ceremony, she was given a deafening send-off with firecrackers exploding all around.

The *Victorious* was bound for the US Naval Base at Yokosuka on the Japanese main island of Honshu, about 90 miles from Tokyo. During the Second World War it had been a major Japanese base and in July 1945 US Navy aircraft had badly damaged the battleship *Nagato* which was berthed there. On passage the *Victorious* steamed through the Strait of Formosa (Taiwan Strait), and as the temperature fell, blue uniforms were taken out of mothballs. On Sunday 10 May, off Okinawa, the *Victorious* rendezvoused with the 80,000-ton USS *Kitty Hawk* for two days of flying exercises but these were cut short because of bad weather and on the morning of Tuesday 12 May she steamed into Yokosuka and berthed alongside Piedmont Pier. The US Navy's hospitality was second to none and

The *Victorious* makes an impressive sight with her ship's company, back in blues again, manning the flight deck for the ship's entry into Yokosuka, Japan, in May 1964. *(Fleet Air Arm Museum)*

A USN S2F (Stoof) Tracker lands on the *Victorious*.

(J. Harper)

among the events they laid on were excursions to Yokohama, Tokyo and Mount Fuji. One intrepid group climbed to the top of the sacred mountain, but most of the ship's company were content to stay in the town of Yokosuka and the Naval Base, where the host ship, USS *Topeka*, offered lavish hospitality and sporting fixtures. A children's party was held and an official cocktail party, and colourful baseball caps were sported on deck as the ship's company bought large quantities from the store at the base, even though they were not yet a fashion item in the UK. At one stage a large notice in one of the US clubs read, 'Congratulations to HMS *Victorious* - in three days you have drunk more beer than the whole US Fleet does in one week.' Obviously the visit was very thirsty work.

It was the morning of Tuesday 19 May when the *Victorious* left Yokosuka and FO2 collected his fleet of Commonwealth ships together in preparation for a short exercise period off Okinawa, after which they set course for Subic Bay. During the passage south the fleet skirted the edge of Typhoon Viola, but they were spared the worst of the weather and when they anchored off Manila on Tuesday 26 May they joined a mighty SEATO fleet of 90 ships representing the USA, UK, Australia and New Zealand for 'Exercise Ligtas', a large-scale amphibious exercise. There was relief all round on Monday 8 June when 'Ligtas' came to an end and after the usual 'wash-up' the *Victorious* returned to the Singapore Naval Base, docking on the morning of Saturday 13 June, having celebrated the first anniversary of the start of the commission. It was virtually the end of the commission for most members of the ship's company and on 17 June they moved into the comparative luxury of HMS *Terror* while the *Victorious* moved into the King George VI dry dock the next day. During the period in dock leave was curtailed somewhat when riots occurred in Singapore City, caused by a racial flare-up between Malays and Chinese. The ship moved out of dry dock on Saturday 18 July and the ship's company moved back on board, after which they spent two days at sea carrying out trials on the rudder following repairs and maintenance work.

It had been decided to recommission the ship by air in two stages, starting in early August when the first flight landed at Paya Lebar Airport and the initial draft left for home. In the space of three weeks almost half of the ship's company were exchanged and during the docking period some 40 men, who had volunteered to stay on for the next commission, were flown home for three weeks' general service leave. On Monday 17 August the *Victorious* sailed from Singapore to carry out her trials and work-up in the local area, before leaving for Australia. Two days into the trials, whilst the Buccaneers of 801 Squadron were carrying out deck landing practice, the undercarriage of one aircraft failed completely and the crash barrier was rigged. However, the crew ejected and they were picked up safely within 15 minutes.

On Wednesday 26 August, in company with the destroyers *Caesar* and *Cavendish*, the *Victorious* left Singapore for Fremantle, Western Australia. The passage was made through the Java Sea and the Lombock Strait, east of Bali, but despite the fact that these were Indonesian waters and that 'Confrontation' was still very much a live issue, the only interest shown in the carrier was by a Badger aircraft. Once into the Indian Ocean the force rendezvoused with the mighty nuclear-powered aircraft carrier USS *Enterprise*, together with her nuclear-powered escorts USS *Long Beach* and USS *Bainbridge*, all of which were on a world tour, and they carried out a short air defence exercise which culminated in a fly-past by the US Navy Phantoms and Crusaders from the *Enterprise*. The *Victorious* arrived alongside No 8 berth at Fremantle on the morning of Tuesday 1 September, and the week spent in the port was much enjoyed by everyone as the ships' companies were well looked after. Those with relatives in other parts of Australia were able to get leave to visit them, and Petty Officer B. A. Martin flew to Sydney to meet his brother whom he had not seen for 16 years. The ship was host to the Australian Minister of Defence and nearly 14,000 members of the public visited the carrier on three open days. The call at Fremantle ended on the morning of Tuesday 8 September, and two days later during flying exercises in the Indian Ocean the rudder jammed again and divers were sent down to carry out an inspection. Although it was made serviceable later that day, the problem recurred the next day and the *Centaur*, *Hampshire*, *Berwick* and *Dido* were sent to escort her back through the hostile waters of the Sunda Strait. The *Centaur* rendezvoused with the *Victorious* in the Java Sea and escorted her to Singapore, where they arrived during the afternoon of Monday 21 September. The *Victorious* went straight back into the King George VI dry dock where she remained for five weeks, during which time an assortment of VIPs, including the C-in-C, came to inspect the troublesome rudder. Whilst the ship was in dry dock the squadrons found themselves assisting with the air defence of Malaysia and the Vixens were put on alert to intercept any intruders. The Gannets flew nightly patrols, while the Buccaneers practised their offensive role and the helicopters were detached into the jungles of Johore to assist the Army and Police in rounding up Indonesian infiltrators.

On Friday 9 October Captain D. L. Davenport OBE RN took command of the ship, and Captain Compston left for promotion to Rear-Admiral and a new appointment in the USA. Captain Davenport joined the Royal Naval College, Dartmouth, as a 13-year-old Naval Cadet in May 1933, and served as a midshipman in the Mediterranean Fleet from 1937 to 1939. Shortly after the outbreak of the Second World War he was Sub Lieutenant of the destroyer *Blanche* when she was mined and sunk in the Thames Estuary in November 1939, the first destroyer to be lost in the war. As a lieutenant he served in the destroyer *Mashona* during the Norwegian campaign and was again a survivor when that ship was bombed and sunk off the Irish coast while returning from the *Bismarck* action in May 1941. As First Lieutenant of the destroyer *Tetcott* he took part in the Allied landings in Sicily, at Salerno and Anzio, and was Mentioned in Dispatches in January 1944. On his return from the Mediterranean he was First Lieutenant of the destroyer *Caesar* in the Home Fleet and took part in convoys to North Russia. During 1945 and 1946 he commanded the frigates *Cotton*, *Holmes* and *Porlock Bay*, the latter on the America and West Indies Station, before serving two years in the cruiser *Sheffield*.

From 1949 to 1951 he served at HMS *Ganges*, the Boys' Training Establishment, and then attended the RN Staff Course before serving with the Indian Navy as Naval Instructor at the Indian Defence Services College. He was promoted Commander in 1951 and awarded the OBE in 1954. On his return from India he commanded the frigate *Virago* and he was the Executive Officer at RNB Chatham. On promotion to Captain in 1957 he served on the staff of the Admiral Commanding Reserves and from 1960 to 1962 he commanded the Inshore Flotilla, Far East Fleet, in HMS *Woodbridge Haven*. Before being appointed to command the *Victorious* he held an administrative post in the Admiralty.

On Friday 30 October the *Victorious* was towed out of the dry dock to No 8 berth, and in early November the second big airlift of the ship's company personnel got under way. Once again a special Movements Office was set up on board as the turnover from the old hands to the new ones posed accommodation problems since the ship was well over strength for at least ten days. However, in mid-November, the last man out of the last trooping flight, M(E) McCormack, was greeted at Paya Lebar by Captain Davenport and on 15 November the recommissioning service was held on board, which included the cutting of a huge cake and a fly-past by the ship's aircraft. Six days later, with the rudder problem having been cured yet again, the *Victorious* left Singapore to work up to full efficiency in local waters and this continued well into December. It had been hoped that she would be able to spend Christmas 1964 in Hong Kong, but with no sign of a let-up in the Indonesian Confrontation she was required to stay within easy reach of Singapore and it was not until Wednesday 23 December that she went alongside the Naval Base for the Christmas and New Year break.

It was on Wednesday 6 January 1965 that the *Victorious* sailed again, this time for exercises with the US Navy in the

HMS *Victorious* alongside No 8 berth at Fremantle, Australia, in September 1964. (J. Harper)

Subic Bay area. During one exercise a Phantom F4 aircraft from USS *Ranger* visited the *Victorious* for touch-and-go landings, and became the first Phantom to 'land' on a British carrier. Later in the month the exercises moved north to Okinawa, and at 5.12pm on Friday 22 January there was tragedy when a Sea Vixen piloted by the commanding officer of 893 Squadron, Lt-Cdr J. A. Sanderson RN, with Lt E. S. Billett RN as the observer, crashed over the ship's bow. Despite a long search no trace was found of the aircraft or the aircrew and two days later a service was held in their memory. During the evening of Tuesday 26 January came a heart-stopping moment for those on watch on the bridge when there was a total rudder failure, but fortunately it was working again within a few seconds. On Friday 29 January the *Victorious* left the Okinawa area and two days later she secured to No 1 buoy in Hong Kong Harbour. It was a welcome break after a very busy month and particularly after the cancelled visit to the city in December 1964, but by Wednesday 17 February the *Victorious* was back in the Subic Bay exercise areas where she rendezvoused with HMS *Eagle* which had just joined the Far East Fleet.

The first two weeks of March 1965 were spent carrying out essential maintenance at Singapore, but by the middle of that month the fleet was gathering off Pulau Langkowi, just off the north-west coast of Malaya near the border of Thailand. Each year the Far East Fleet carried out Flag Officer's Training Exercises which were known as 'Fotex', and they were usually held off the east coast of Malaya in the South China Sea, but with 'Confrontation' continuing it was decided to transfer the location to the Andaman Sea and the Strait of Malacca. Four aircraft carriers took part in the exercise - *Eagle*, *Bulwark*, *Victorious* and HMAS *Melbourne* together with 36 other ships of the Royal and Commonwealth Navies. It was a formidable show of strength which signalled British and Malaysian determination to defend the new Federation of Malaysia. There was tragedy during the exercise when one of the *Eagle's* Wessex helicopters ditched with the loss of two of its three crew members, but for many the highlight of the 12 days were the banyans on Pulau Langkowi, which is an exclusive holiday resort today. On Saturday 27 March, as

The *Victorious* alongside the Naval Base at Singapore. (*J. Harper*)

the fleet returned to Singapore, the Prime Minister of Malaysia, Tungku Abdul Rahman, together with ministers and defence officials, embarked in the *Eagle* for the highly publicized 'Exercise Showpiece', flying displays by the RAF and Fleet Air Arm Squadrons, which were as much for President Sukarno's benefit as the other observers. That evening the whole fleet berthed at Singapore's Naval Base. Next day the *Victorious* was at sea once again, this time for her Operational Readiness Inspection, which was concluded with a weekend at Pulau Tioman. After leaving the tropical island on Monday 5 April, the *Victorious* set course for Kobe which was reached seven days later. Not only were the ship's company enthusiastic about visiting Japan, but the local people were also keen to see the carrier and on Maundy Thursday 7,000 people were able to go on board. However, on Easter Saturday there were even more people waiting on the jetty over two hours before the gangways opened, and Captain Davenport decided to let them start coming aboard. As the day went by more and more people turned up until it was estimated that there were well over 15,000 waiting their turn, and in the event the gangway had to be closed, leaving thousands of disappointed would-be visitors. After leaving Kobe on the morning of Monday 19 April the ship made a 200-mile passage of the Inland Sea, before carrying out flying exercises off Okinawa. It was during this period, at 12.05pm on Wednesday 21 April, as a Sea Vixen was being launched, its starboard engine caught fire and it ditched about a mile ahead of the carrier in a position Lat 25° - 29'N/Long 127° - 48'E. The ship immediately went full astern and made an emergency turn to avoid the wrecked aircraft, whilst the SAR helicopter made straight for the scene. Within five minutes of the aircraft ditching, the SAR diver was in the water and getting the pilot winched to safety. Sadly, the observer, Sub-Lt Rainsbury RN, had been killed in the accident and his body was also brought on board. Both men had ejected from the aircraft, but only the pilot survived. That same evening, at 6.40pm, the funeral service was held for Sub-Lt Rainsbury and his body was committed to the deep. Three days later the *Victorious* moored in Hong Kong Harbour.

After two weeks in the colony very few had much money left, but looking ahead the ship's deployment east of Suez was coming to an end and after leaving Hong Kong the *Victorious* steamed to Manila to take part in the final SEATO exercise of the commission, the escorting of a military convoy from Manila to Bangkok in the face of submarine and air opposition. Also taking part in 'Exercise Seahorse' were HMAS *Melbourne* and USS *Bennington*, together with HM Ships *Agincourt*, *Chichester*, *Plymouth* and HMAS *Parramatta*. The exercise finished on Saturday 22 May in the Gulf of Thailand and the escort ships were able to steam upriver to enjoy the delights of Bangkok, but the three carriers had to anchor off Bang Saen and only privileged leave was granted. Just a few were able to make the four-mile boat ride and 60-mile bus trip to see the sights of the city, although they were thoroughly soaked in a tropical downpour on the way. By Wednesday 26 May the *Victorious* was back in Singapore and in June she took part in 'Exercise Windy Weather' in the Sarawak area and off the coast of Malaya, and finally on Saturday 19 June she secured alongside No 8 berth at the Naval Base for the last time in the commission.

At 11am on Wednesday 30 June, with her paying-off pennant flying proudly, the *Victorious* left Singapore, bound for Aden and home, after almost two years service

'Exercise Fotex 65' took place in the Andaman Sea. In this view looking astern from the *Victorious* are HMAS *Melbourne* and HMS *Bulwark*. (J. Harper)

with the Far East Fleet. However, before she left the Strait of Singapore the ship was called to 'Action Stations' when Indonesian destroyers were sighted acting suspiciously near the island of Sambu. Later that afternoon, as the Indonesian warships disappeared, the ship's company were stood down. During the voyage across the Indian Ocean a slightly southerly course was steered and the ship crossed the equator, with the traditional ceremony being held on the morning of Wednesday 7 July. Next day she rendezvoused with the *Ark Royal* which was relieving her on the station, and the squadrons made a dramatic fly-past over the *Ark*, before the *Victorious* set course for Aden, where she arrived four days later. By mid-1965 the colony was gripped in a terrorist campaign by the NLF and

FLOSY, and it was a great relief to leave after 48 hours and set course for Suez. The carrier made her northward transit of the Suez Canal on Sunday 18 July and seven days later she launched her squadrons to their respective bases, before anchoring at Spithead on the morning of Tuesday 27 July. At 11am that morning, just half an hour before she weighed anchor to steam up-harbour, the ship's company manned ship to cheer Her Majesty The Queen as she returned by hovercraft from a visit to the Isle of Wight. At 12.15pm that day the *Victorious* secured alongside Middle Slip Jetty, and the commission was over. During her two-year absence she had steamed 125,000 nautical miles, and had notched up well over 10,000 landings on her flight deck. It was an impressive record for an old lady.

On Saturday 27 March 1965 the ships of the Far East Station took part in 'Exercise Showpiece', staged for the Prime Minister of Malaysia. In this view HMAS *Yarra* leads a steam-past.

(*J. Harper*)

Chapter Fourteen

The Final Commission

Following her return to Portsmouth in July 1965, the *Victorious* was moved into D Lock for dry docking so that her rather unreliable rudder could be removed for a thorough overhaul, and in the chilly damp days of November 1965 the ship's company began to assemble for the carrier's eighth and, as it transpired, last commission. However, it was in the spring of 1966 that the refit was completed and on the afternoon of Thursday 7 April the Commissioning Ceremony was held while the ship was berthed at Middle Slip Jetty. First to arrive for the service were the families and friends of the ship's company followed three-quarters of an hour later, at 2.45pm, by the VIPs. In addition to the C-in-C Home Fleet, Admiral Sir John Frewen, both the C-in-C Portsmouth, Admiral Sir Frank Hopkins, and the FOAC, Rear-Admiral William O'Brien, also attended the ceremony, as did the Lord Mayor and Lady Mayoress of Portsmouth. The service was conducted by the Chaplain of the Fleet and following this Captain Davenport addressed the ship's company. Then the 1,000 or so relatives who attended the ceremony were served tea in B hangar and, in keeping with tradition, Mrs J. Davenport, the captain's wife, cut the Commissioning Cake, ably assisted by the youngest member of the ship's company. Five days later, on the afternoon of Tuesday 12 April, the *Victorious* left Portsmouth for her sea trials in the Lyme Bay and Portland areas, where the Wessex helicopters

The *Victorious* in D Lock, Portsmouth Dockyard. The uncompleted aircraft carrier *Leviathan* can be seen in the background moored in Fareham Creek. *(Lt N. Curnow)*

Being towed stern first out of D Lock. (Lt N. Curnow)

of 814 Squadron joined, and after a long weekend at Portsmouth the carrier sailed for the grey waters of the Moray Firth in company with the frigate *Blackwood*. Over the next ten days she carried out flying trials with the Vixens of 893 Squadron, and the improved Buccaneer Mk IIs of 801 Squadron, before returning to Portsmouth on Tuesday 24 May to find the deep water channels obstructed by dredgers which meant she had to anchor at Spithead until the early hours of the following morning, and it was almost 5am before she secured alongside Middle Slip Jetty. On the last day of May the *Victorious* put to sea for her second work-up period and on Monday 6 June the Minister of Defence (Navy), Mr J. P. W. Mallielieu MP, visited the ship by helicopter. Three days later, during the late afternoon of Thursday 9 June, while the ship was launching aircraft in foggy conditions off Lizard Point, Buccaneer 232 crashed off the port bow following take-off. Fortunately, the crew were rescued by helicopter and naval divers working from the diving tender *Reclaim* recovered the wreckage. As it was the first Mk II Buccaneer to crash,

MoD experts were especially anxious to find the cause of the accident. On the morning after this incident, with her work-up completed, the *Victorious* returned to Portsmouth where the ship's company could take some leave and the carrier could be prepared for her departure to join the Far East Fleet once again.

The *Victorious* left Portsmouth on the afternoon tide of Friday 8 July 1966 when, on the order 'Let go head rope', a group of past *Victorious* officers slipped the moorings and she left Middle Slip Jetty in company with the frigate *Galatea* to steam through the Solent and into the Channel where all the squadrons were recovered. There was only a brief halt in Gibraltar Bay, where the first mail from home was delivered, before the ship resumed her eastward course for the exercise areas off Malta where the final phase of her work-up was completed and her Operational Readiness Inspection was carried out. It was during this period, on 19 July, that an emergency signal was received from the small Italian island of Lampedusa, situated approximately halfway between Malta and the North African coast of

HMS *Victorious* leaves Portsmouth on Friday 8 July 1966. HMS *Dolphin* and the Gosport ferry terminal can be seen in the background. (*Lt N. Curnow*)

Tunisia, where a woman was seriously ill during childbirth. A doctor was immediately flown to the island in a helicopter of 814 Squadron, and owing to the seriousness of the woman's condition, she was quickly embarked in the Wessex and with the doctor flown to Bighi Hospital in Malta. Sadly it was of no avail, for on arrival at hospital both the mother and child were found to be dead, but the following message was received from the Mayor of Lampedusa: 'Deeply moved and impressed by high sense of humanity shown in prompt air assistance to the sick woman Perrone. I wish to express the profound admiration and gratitude felt by myself and the whole population. Your spontaneous action confirms once again your proverbial compassion towards human suffering transcending all national barriers.'

Next day the *Victorious* steamed into Grand Harbour for the first 'foreign' run ashore, but it was only a brief visit and on Sunday 24 July, despite a strike by tugmen, the carrier left harbour and set course for Cyprus where flying exercises were carried out with the RAF. However, after being flown ashore to collect the mail, the COD Gannet refused to leave RAF Akrotiri and during the evening of 26 July the ship anchored off the coast so that the plane could be returned on board by lighter, together with the mail. Next afternoon the carrier anchored off Port Said and on the morning of Thursday 28 July she made her southbound transit of the Suez Canal as the 17th ship in a convoy of 19. For many it was their first glimpse of the 'mysterious east', and in those days before wide-bodied jets and 'plastic' airports brought mass tourism to the most remote parts of the globe within 24 hours, the passage through the Suez Canal was always very exciting. Goofers crowded the flight deck and the Gully-Gully men baffled everyone by producing their small, yellow, bewildered chicks from the most unlikely places. Late afternoon found the ship in the Great Bitter Lake where the northbound convoy was waiting to continue its voyage, among them the minesweeper HMS *Calton* which exchanged salutes and good wishes. It was 8pm that evening when the *Victorious* cleared the Suez Canal and set course into the darkness of the Gulf of Suez for Aden. As she steamed through the Red Sea and the air temperature rose, the air-conditioning decided to stop working and temperatures on board rose to 140°F in some places. On the morning of Sunday 31 July the ship anchored in the outer harbour of Aden where mail was embarked and some sports parties were landed. However, since her last call at the port 12 months earlier, the terrorist incidents in Aden State had increased alarmingly with mortars and rocket launchers being used, and only the day before a number of Mig fighters had crossed over the border from Yemen and attacked the home of a local ruler. There were no regrets when later in the day the carrier weighed anchor and put to sea. Next morning, at 9.35am, the C-in-C Middle East, Admiral Sir Michael Le Fanu, together with the Defence Minister of Aden State, the Sultan Saleh, flew on by helicopter for a six-hour visit, and the British Military Attaché for Khartoum in the Sudan, together with his two tons of baggage, was ferried ashore. After leaving the Aden area the ship set course for Gan, and en route she passed within 60 miles of HMS *Eagle* which was steaming north from Mombasa, and at this stage the *Victorious* officially took over from her as the operational strike carrier east of Suez. One nice personal touch while the two ships were in the area was when a junior rating on board was able to speak by radio to his father who was serving in the *Eagle*. As the ship steamed towards Gan some members of the ship's company were given flights round the ship by the helicopters of 814 Squadron, and off Gan another rating was able to visit his father in RFA *Wave Knight*. During the late morning of Friday 5 August, whilst operating aircraft in the area off Gan, Vixen 246 ditched shortly after launching but, fortunately, the crew were rescued by the SAR helicopter. During these operations the equator was crossed and recrossed on a number of occasions, but it was decided to postpone the traditional ceremony until the ship's schedule was less busy and on Sunday 7 August course was set for the Strait of Malacca. On Thursday 11 August the *Victorious* arrived at the northern end of the strait and for four days she carried out flying exercises off the island of Penang, with the Buccaneers of 801 Squadron leaving the ship for RAF Changi on Singapore Island (now Singapore's main civilian airport). There was even time for banyans at Pulau Langkowi, and at 8.30am on Tuesday 16 August the carrier secured alongside No 8 berth at Singapore Naval Base, marking the end of the first phase of the foreign leg of the commission.

During her three-week maintenance period at Singapore there was plenty of time for relaxation and the ship's sports teams made their mark on the playing fields and in the swimming pool of HMS *Terror*, while others enjoyed a few days in the cool air of the hill resort of Fraser's Hill. Nevertheless, there was still plenty of work to be done and a reminder of this came on Monday 22 August when FO2 Far East Station, Rear-Admiral Charles P. Mills CB CBE DSC, hoisted his flag in the *Victorious* and on the morning of Monday 5 September the carrier left Singapore in company with HM Ships *Hampshire* and *Leander* and HMAS *Parramatta* to carry out flying exercises in the Singapore and Subic Bay areas. During the passage to the Philippines an RAF Shackleton from Changi made a mail drop, and a three-ship RAS was carried out. On Monday 12 September there was a two-day break when the ship went alongside Alava Pier at the US Naval Base at Subic Bay, where the dubious pleasures of Olongapo were sampled by some. After sailing from Subic Bay exercises were continued in the area, during which Vixen 247 was lost but, fortunately, the crew were recovered safely. On

Another view of the *Victorious* as she leaves Portsmouth on 8 July 1966.

(*Lt N. Curnow*)

The *Victorious* arrives at Malta on 20 July 1966. (*Lt N. Curnow*)

The view from the bridge as the *Victorious* steams south through the Suez Canal on Thursday 28 July 1966.

(*Lt N. Curnow*)

Saturday 17 September the 1,000th deck landing of the commission was recorded, and a few days later course was set for Hong Kong. It was during what turned out to be a rough passage that a signal for assistance was received from a merchantman, the SS *August Moon*, which had run aground on Pratas Reef in the South China Sea, but in the event the USS *Oriskany* arrived on the scene first and so the *Victorious* continued her voyage and tied up to No 1 buoy in Hong Kong Harbour at just before 1pm on Friday 23 September. Next day, at 10.30am, the ship's new commanding officer, Captain I. S. McIntosh DSO MBE DSC RN, arrived on board and an hour and a half later he assumed command from Captain Davenport who left the ship for promotion to Rear-Admiral and an appointment in the Admiralty.

Captain McIntosh was born and brought up in Geelong, Victoria, Australia, and from there, at the age of 18, he entered the Royal Navy. During 1939 and 1940 he served as a midshipman in the Mediterranean, East Indies, South Atlantic and in northern waters, before volunteering to serve in submarines. In March 1941 he had been a passenger in the 8,800-ton Anchor Line steamer *Britannia*, which was en route to India, although Sub-Lt McIntosh was bound for Alexandria to join the Submarine Flotilla based there. Being wartime, with the Mediterranean closed, the vessel was making the long voyage by way of Cape Town. On the morning of Tuesday 25 March, when the *Britannia* was in the Atlantic some 600 miles off Freetown, Sierra Leone, she encountered the German commerce raider, *Thor*, which shelled the liner for an hour, severely damaging her. When the order to abandon ship came, Sub-Lt McIntosh manned No 7 lifeboat which had been damaged in the action, and in the end there were 82 people in the boat which was designed for 56. During the night boat 7 became separated from the other lifeboats and Sub-Lt McIntosh succeeded in carrying out makeshift repairs to the damaged boat, by hanging over the side in order to do so. They tried to sail eastward towards Africa, but could

A Buccaneer flies over the *Victorious*. (Lt N. Curnow)

HMS *Victorious* steams through the Johore Strait. (Lt N. Curnow)

A Gannet is launched...

...and then a Buccaneer.
(Lt N. Curnow)

make no progress against the north-easterly winds and so decided to make for Brazil instead. There then followed a gruelling 1,600-mile voyage across the Atlantic, with Sub-Lt McIntosh and Third Officer McVicar the only real seamen on board, through storms and searing heat, until after 23 days they landed near Sao Luis in Brazil. Only 38 survivors were left, and for his part in the epic voyage Sub-Lt McIntosh was awarded the MBE. When he arrived back in the UK Lt McIntosh joined the submarine *Porpoise* and in 1942 he was the First Lieutenant of the *Thrasher* in the Mediterranean, where he was awarded the DSC. Returning to the UK later in 1942 he took command of *H44* and later the *Sceptre*, which operated off the coast of Norway and towed the midget submarine *X10* to attack the *Tirpitz*. For these exploits he was awarded the DSO and he was twice Mentioned in Dispatches. In 1944 he went ashore to help in the development of submarine radar, and after the war he commanded the submarines *Alderney* and *Aeneas*, and instructed submarine COs. From 1956 to 1958 he was the Commander of HMS *Ark Royal*, during that carrier's second commission, and before taking command of the *Victorious* he commanded the 2nd Submarine Squadron and was the Deputy Director of the Naval Tactical and Weapons Policy Division.

Hong Kong provided many pastimes and attractions for the ship's company, from the beaches of Repulse Bay to the bars of Wanchai and Kowloon, and FO2 entertained Mr Mallalieu, the Minister of Defence for the Navy. It was here that the final of the Beard Growing Competition took place and it was judged by FO2 and his more popular assistants, two young actresses. Finally, at 10.30am on Tuesday 4 October the *Victorious* sailed from Hong Kong for Subic Bay to carry out more exercises in the area, and it was at 9pm, just after darkness had fallen, on Friday 7 October that a helicopter of 814 Squadron crashed into the sea about a mile astern of the ship. The planeguard frigate, HMS *Cleopatra*, quickly closed the scene which was illuminated with flares, and the seaboat was lowered, but despite an intensive search which lasted well into the early hours of the following morning, the four members of the crew, Lt-Cdr Neil Whitwam, Lt-Cdr Alan Johnson, Sub-Lt Paul Dyer and Ldg Seaman John Gee, could not be found and were presumed dead. On conclusion of the exercises, at 4.30pm on Monday 10 October, as the ship steamed through the San Bernardino Strait, between the main Philippine island of Luzon and the island of Samar, a memorial service was held on board to pay respects to the lost aircrew.

The next deployments of the commission were the eagerly awaited visits to Sydney and Fremantle in Australia, and at 12.40pm on Saturday 15 October the ship crossed the equator at Longitude 140° - 15'E, and this time there was a full ceremony involving King Neptune and all his court. However, it was soon back to business when, on Tuesday 18 October, the *Victorious* participated in the most extensive exercise of the commission, 'Swordhilt', which took place down the eastern seaboard of Australia and involved 23 ships of the Royal Navy, the Royal Australian Navy, including HMAS *Melbourne*, and the US Navy. During the exercise the *Melbourne* effected an unusual rescue of one of the *Victorious*' helicopters when it developed rotor trouble causing it to remain hovering about 30 feet above the water without being able to move sideways. The *Melbourne* altered course and steamed to a position directly beneath the helicopter so that it could land, and while the Australian carrier was approaching, two members of the helicopter's crew jumped into the sea in case it crashed on deck. The pilot, Lt Michael Sant, stayed in the helicopter and landed safely as the *Melbourne* positioned herself underneath. During the afternoon of Thursday 27 October the squadrons from both the *Victorious* and *Melbourne* made a historical fly-past over Sydney, with two Admirals amongst the aircrew, Vice-Admiral Mills and Vice-Admiral McNicol, Chief of the Australian Naval Staff. This led to the following signal being sent to the Ministry of Defence: 'Believe this first occasion so much Vice airborne in one flight.' Next morning the *Victorious* entered Sydney Harbour and secured alongside her berth at Woolloomooloo. Accompanying the carrier during her visit to the Australian ports were the guided missile destroyer HMS *Hampshire*, the frigates *Arethusa*, *Cleopatra* and *Leander* and the submarine *Oberon*, and when the ships were opened to the public in Sydney Harbour, over 15,000 people visited them. As for the ships' companies, they enjoyed as always the extraordinary generosity of their Australian hosts. On board the *Victorious* there were memories of a bygone age when, under the inspiration of the commanding officer of 814 squadron, Lt-Cdr Michael Apps, a Korean War Firefly was purchased from RANAS Nowra for the passage home and eventual presentation to the Fleet Air Arm Museum at RNAS Yeovilton. While in the port many men took the opportunity to visit relatives and when the *Victorious* left Sydney on the morning of Thursday 10 November, one intrepid group set off for Fremantle in the ship's Landrovers, arriving there one day ahead of the carrier. The *Victorious* faced an eight-day voyage across the Great Australian Bight after which she arrived alongside No 7 North Wharf, Fremantle, at 9am on Friday 18 November to a wonderful welcome from the local population. Once again thousands of people queued to view the ship, which hosted yet another children's party, but probably the most popular visitor on this occasion was 'Miss Western Australia' who helped to stir the Christmas pudding mixture. It was to be the ship's last visit to Australian shores and at 9am on Friday 25 November, to the fond farewells of many friends, she left Fremantle and set course for the Naval Base at Singapore. During the passage north, whilst

HMS *Victorious* at Hong Kong with Kowloon in the background...

...and with the more familiar skyscrapers of Victoria on Hong Kong Island forming the backdrop.
(Lt N. Curnow)

The *Victorious* berthed at Woolloomooloo, Sydney, in November 1966. (Lt N. Curnow)

passing the Cocos Islands, a signal was received requesting the services of a Chaplain to celebrate Holy Communion and to baptize a baby. The Chaplain from the *Victorious* was landed by helicopter and was, in fact, the first clergyman to visit the island since *Ark Royal* had passed the islands earlier in the year. As the ship passed Christmas Island the Vixens staged a fly-past over the island and during the passage a five-ship RAS was accomplished. After steaming through the Sunda Strait, the *Victorious* arrived back at Singapore Naval Base on the morning of Friday 9 December to carry out a Dockyard Assisted Maintenance Period, and to celebrate Christmas. Despite the tropical heat the festivities were observed in the traditional manner and there was a good turnout, with over 1,000 attending from the Naval Base, for the Christmas carol concert on the flight deck. All the mess decks were decorated festively and the electrical department adorned the starboard side of the island with Christmas lights. On Christmas Day Captain McIntosh visited the mess decks and the sickbay and in the following week the ship staged a pantomime, 'Vic Whittington', at HMS *Terror*. The New Year was rung in by the youngest member of the ship's company, JS T. J. Addington, and despite the hangovers, work went on to ensure that the ship was made ready for sea once again.

On the morning of Wednesday 4 January 1967 the *Victorious* left Singapore to begin the second half of her foreign leg, and after embarking her squadrons she began flying exercises in the South China Sea, off Pulau Tioman. On Friday 6 January the C-in-C Far East Station, Vice-Admiral Sir Frank Twiss, and the FOAC, Rear-Admiral W. O'Brien, visited the ship, although only the former left at the end of the day, with the latter staying for five days before transferring to the *Bulwark* which had just returned from a mercy mission to a Yugoslavian steamer aground in the South China Sea. Soon after the departure of FOAC the *Victorious* set course for Hong Kong, and she arrived in the harbour during the afternoon of Monday 16 January, where the cold weather necessitated a change back into

Leaving Sydney on 10 November 1966, with the Harbour Bridge in the background. *(Fleet Air Arm Museum)*

170

A Sea Vixen on touch-and-go circuits...

...and a Gannet approaches for a perfect landing. *(Lt N. Curnow)*

Sea Vixens fly past in formation over Christmas Island.
(Lt N. Curnow)

blue uniforms. Five days into the visit the ship's company were entertained by the singer Anne Shelton who gave a concert in the hangar, and on Wednesday 25 January the ship left for the Subic Bay exercise areas. Two days later, at 7.13pm on 27 January, there was another fatal accident when Vixen 242 crashed into the sea off the starboard bow shortly after launching. Although the pilot of the aircraft, Lt A. C. Selman was rescued, the observer, Lt R. A. Brown was lost, despite a search which went on well into the following day. The first weekend in February was spent in the US Naval Base at Subic Bay, and flying resumed when the carrier sailed again on the morning of Monday 6 February. Nine days later, at 4am on Wednesday 15 February, a distress signal was received from a 1,700-ton Thai-registered coastal steamer, *Maha Thevi*, which had gone aground on the tiny, rocky island of Permanggil, 75 miles north-east of Singapore. When she received the signal the *Victorious* was some 85 miles away and she steamed at 28 knots to reach the scene as quickly as possible. There were over 40 people on board the coaster which was being pounded on the rocks by heavy seas, and two of the crew had drowned whilst trying to reach the shore. Despite the severe weather, at 6.35am a helicopter was launched to fly to the stricken vessel where the observer was lowered to inspect the damage and report on the assistance which was most needed. As a result two officers from HMAS *Vampire*, which was also standing by, were lowered onto the *Maha Thevi* and they organized a bosun's chair line to the shore. This enabled the crew to be landed safely, and next morning between 7am and 9am the helicopters of 814 Squadron provided a shuttle service between the ship and the island, so that the survivors could all be embarked in the *Victorious* for the passage to Singapore, where the carrier arrived on the morning of 17 February.

The next two weeks were spent carrying out essential maintenance and preparing for FO2's annual training exercise, in this case 'Fotex 67', which included the frigates *Falmouth*, *Brighton*, *Llandaff* and *Londonderry*. As always the most popular parts of the exercise were the weekend breaks at Pulau Tioman (5 March), and Pulau Langkowi (11 March). At this point one officer, with the return to Portsmouth fast approaching, decided it was time to rid himself of his dilapidated Ford Consul motor car and rather than send it to the scrapyard he had it duly launched from the port catapult. The exercises ended on Friday 17 March, when the fleet returned to Singapore, and eight days later the *Victorious* was at sea once again for flying exercises in the Subic Bay area, followed by a visit to Hong Kong where she arrived on Monday 3 April. During the stay, on a very foggy day, one of the famous Star Ferries, *Mao Tin*, collided with the starboard side of the carrier whilst she was at her mooring. Luckily, there was no real damage to either vessel and two days later, on 11 April, the *Victorious* made her final departure from Hong Kong. During the passage back to Singapore there were more exercises off Subic Bay and on 24 April she returned to the

Naval Base at Singapore. On Thursday 27 April Rear-Admiral E. B. Ashmore relieved Vice-Admiral Mills as FO2 Far East Station, but the new Flag Officer would not fly his flag for long in the *Victorious*, for at 9.30am on Thursday 4 May, with her paying-off pennant flying, she left Singapore for the last time, bound for Aden. Nine days later she anchored off Aden's Outer Harbour, where the terrorist campaign had reached new heights and service families were being evacuated, but the ship's company were still able to find time for swimming and shopping in a heavily guarded Steamer Point. On Sunday 21 May 1967, only 15 days before the start of the so-called Arab-Israeli Six-Day War, the *Victorious* made her northbound transit of the Suez Canal and set course for Malta. Everywhere on board there was a sense of anticipation as the ship was due in Portsmouth on 1 June and there were now only ten days to go. Once through the Suez Canal the atmosphere on board became noticeably more relaxed, but unfortunately with political tensions mounting in the Middle East between Israel and her Arab neighbours, the Government decided that the *Victorious* was required to remain in the Mediterranean until the crisis had been resolved, and during the passage from Port Said to Malta Captain McIntosh made the announcement that the return to Portsmouth was being delayed. The news caused a certain amount of irritation among families at home in the UK and some complaints were widely reported in the national press. However, 11 men who had made arrangements to get married upon their return were given special permission to fly home and they arrived in Lyneham in early June.

Three days after leaving Port Said the *Victorious* secured alongside Parlatorio Wharf in Malta's Grand Harbour on the morning of Wednesday 24 May, but only for long enough to refuel and that afternoon she returned to sea with HMS *Rhyl*, remaining in the exercise areas off the island. On 30 May she returned to Grand Harbour, but three days later she was at sea again, still in the vicinity of Malta and being closely watched by a Soviet warship. On the morning that the Six-Day War began the *Victorious* returned to Grand Harbour and she remained at No 11 buoy for seven days until Monday 12 June when she left the island to resume her homeward passage. The decision to release the *Victorious* was a clear indication that, as far as Britain was concerned, there would be no involvement in this Middle East crisis. On Tuesday 13 June the remaining jets were flown off to their stations and on the following day the ship made a 48-hour stopover in Gibraltar, just long enough for ammunition packing cases to be embarked. When she resumed her voyage she had one additional passenger on board in the form of nine-year-old Sheila Wareham, the daughter of an Army NCO, who was returning home for an operation on her ear and the sea voyage was necessary to avoid any strain which air travel might cause.

On Monday 19 June the Gannets and the helicopters were flown off and that afternoon the ship secured to C buoy in Plymouth Sound. As well as clearing Customs, 814 Squadron's ground crews were disembarked, and next day at 8pm the ship left Plymouth to steam up-Channel for Portsmouth. Next morning, at Spithead, 300 relatives were embarked by tug and they accompanied the ship into harbour where, at 12.30pm, she secured alongside Middle Slip Jetty and another 500 eager relatives streamed up the gangway. The *Victorious* was home at last, but unknown to anyone at the time she would never put to sea under her own power again.

A five-ship RAS in the Java Sea, November 1966, with the *Victorious* in the centre. (*Fleet Air Arm Museum*)

HMS *Victorious* floodlit at Singapore Naval Base. (*Lt N. Curnow*)

'Exercise Fotex 67'. Ships of the Far East Fleet anchored off Pulau Langkowi. HMS *Bulwark* is anchored ahead of the *Victorious* and the frigates include HM Ships *Brighton, Falmouth, Llandaff* and *Londonderry*.
(Lt N. Curnow)

The *Victorious*, with her paying-off pennant flying, leaves Singapore for the last time on the morning of Thursday 4 May 1967.
(Lt N. Curnow)

Chapter Fifteen

The End Of The Story

As always at the end of a commission the *Victorious* underwent a refit and she was due to recommission on 24 November 1967 for what was generally accepted would be the old lady's final operational service. In fact, in the autumn of 1965 the Government had considered the possibility of buying an Essex-class aircraft carrier from the USA, to act as a stopgap until the Royal Navy's new aircraft carrier CVA 01 became operational and to replace the *Victorious*. However, in February 1966 the Government's Defence Review killed off CVA 01 and put forward the idea of a complete British withdrawal from its commitments east of Suez. Although there was a great deal of opposition to the idea, with Britain's post-war role as a world power greatly diminished it was inevitable, and after centuries of conflict with her European neighbours it was now desirable that she should turn instead towards Europe, and for all future British Governments, whatever their shade of political opinion, Europe would take the place of the Commonwealth in foreign affairs. However, the Labour Government of the 1960s made the mistake of deciding that a withdrawal from east of Suez would mean that aircraft carriers were no longer required as a weapon of defence, and the decision to scrap all the Royal Navy's carriers was made easier when the short-term reduction in defence costs was calculated. Politicians, who are always looking for cuts in expenditure, almost inevitably turn to the Defence budget to find them, and the *Victorious*, because of her age, presented an easy target. But all this was still in the future, and on Tuesday 8 August 1967 the *Victorious* was towed round from Middle Slip Jetty to D Lock for dry docking where her troublesome rudder was to be completely overhauled. The new commission actually began in August

Looking very forlorn the *Victorious* is pictured here laid up in Fareham Creek on 27 June 1969. Just a few days after this photograph was taken it was announced that she was to be sold for scrap. *(M. Lennon)*

With her large radar dome gone, and her flight deck deserted, the *Victorious* is towed from Portsmouth Harbour during the afternoon of 11 July 1969 for her final destination, the shipbreaker's yard at Gare Loch.
(*The News, Portsmouth*)

and her ship's company were accommodated in HMS *Centaur*, the light fleet carrier which had ended her operational service in August 1965 and had been put to use as an accommodation ship ever since.

Fire at sea is one of the most dangerous hazards for any ship, and when a vessel is laid up in dry dock without its own power and services the danger is even greater and so fire precautions and patrols are very stringent. On Thursday 24 August there had been a fire alarm during the forenoon, but this turned out to be nothing more than a minor electrical fault. On Monday 11 September the fire alarms again sounded during the forenoon watch as a fire was reported in 6Q flat. This time both the Dockyard and Portsmouth City Fire Brigades attended, but once again it was nothing very serious. Later that same day, at just before midnight, an electrical fire was reported on a mess deck, and both the Dockyard and City Fire Brigades attended again, despite the fact that it was extinguished within minutes and the request for the services had been cancelled. During October there were five fire alarms, all of them in working hours, but with the ship swarming with dockyard workmen using welding equipment which generates a lot of smoke, it is not surprising that most of them were false alarms. On Friday 20 October D Lock was flooded, but the dockyard firemain system remained connected as the ship's machinery was still shut down and unable to supply this vital function. Eleven days later, at 8am on Tuesday 31 October, the ship's company began their move back on board from HMS *Centaur* and later that afternoon, with the transfer completed, everyone began to settle into their new home. On the morning of Friday 3 November lower deck was cleared and Captain McIntosh addressed the ship's company for the first time, and on Thursday 9 November FOAC, Rear-Admiral L. D. Empson, visited Captain McIntosh. No doubt they discussed the progress of the refit as the day approached for the formal recommissioning ceremony. They probably also discussed the routine change of command which was due to take place in June 1968 and the forthcoming commission which would take the carrier east of Suez once again in the early months of 1968. Everything seemed to be going according to schedule and the ship would soon be at sea again.

At 6am on Saturday 11 November 1967, Chief Petty Officer T. Catterson, who was sleeping in No 13 Mess on the starboard side of Two Deck in K compartment, suddenly awoke to smell burning and to see smoke coming from a vent above the mess doorway. Within minutes he had raised the alarm and the ship's fire party were tackling the blaze with fire extinguishers. Unfortunately, the ship's firemain system failed and they were soon forced back by the dense smoke, but according to the ship's log, the Dockyard and City Fire Brigades were fighting the blaze by 6.20am. From outside the ship there was little sign that anything was wrong, but inside the firefighters were hampered by dense smoke and the fact that there were no lights. At 7.20am Commander G. C. Lloyd, the ship's new Executive Officer, was informed and both he and the Admiral Superintendent of the Dockyard arrived an hour later. In the event it was 11.30am before the fire was extinguished, and it was soon realized that Chief Petty Officer J. C. Nicol, who had also been asleep in the mess deck, had been trapped and had tragically lost his life in the fire. A subsequent Inquiry found that the cause of the blaze was a hot water tea urn which had been left switched on overnight, and which had boiled dry. Having no automatic cut-out, the overheating element had set fire to the insulating wire and this had quickly spread to surrounding do-it-yourself material which had been used to make improvements to the bar area of the mess. That afternoon men of the weekend duty watches started to clear up the mess. Although the fire had been tragic and was a setback to the progress of the refit, and despite the fact that the dockyard would have to replace buckled bulkheads, there was no doubt that the work could still be completed on time. On Sunday 12 November the C-in-C Western Fleet, Admiral Sir John Bush, spent an hour with Captain McIntosh and after the conference the commanding officer made a statement to the press: 'We shall recommission on 24 November and we have plenty of time to put things right before sea trials begin in February. We can carry on with crew training in the dockyard while repairs are made.' As to the cost of repairs he said: 'Our radar, missile system, catapults and landing gear are undamaged and unless our time in dockyard hands has to be extended I should have thought that £1 million was too high.' Most significantly, in commenting on remarks about the damage which the MoD had issued to the press he said: 'They were reported as saying that there was "indescribable chaos" on board. This is so untrue that it is not even funny. If it were true then I ought to be sacked.' It would soon become clear that the fire, however serious, was to be used by the politicians for their own ends.

In November 1967 Harold Wilson's Government was facing the worst financial crisis the country had seen for almost 20 years, and the Chancellor, James Callaghan, had devalued the pound by reducing the exchange rate from $2.80 to $2.40. This was coupled with an increase in interest rates and the moves indicated a complete reversal of Government policy. Today those times are best remembered for Harold Wilson's television broadcast when he announced that the devaluation, '...does not mean, of course, that the pound here in your pocket or purse or in your bank has been devalued.' It was in this atmosphere of crisis that the Government was looking for cuts in public expenditure and in the days following the fire rumours about the *Victorious* being decommissioned and scrapped were rife. Nevertheless, the ship's company and the dockyard employees worked hard to prepare the ship for

A last view of the once proud carrier as she is towed into the Solent. *(Maritime Photo Library)*

sea, and during the afternoon of Tuesday 21 November Commander Lloyd addressed the ship's company to confirm that the Commissioning Ceremony would go ahead as planned on 24 November. However, next day, Captain McIntosh was officially informed that the *Victorious* would not recommission and at 4.10pm he cleared lower deck and told the ship's company the sad news, which appeared in all the newspapers on the following day, but he was also able to announce that the Recommissioning Ceremony would go ahead as planned but the occasion would be a 'Families Day' instead.

It was said that there was an air of unreality on board when the ceremony took place, with the ship looking spick and span, neatly painted and buzzing with activity, but about to be discarded. On display was a large commissioning cake which was the pride of the catering department, and five-year-old Alison Taylor, the daughter of PO Kenneth Taylor, presented a bouquet to Mrs Elizabeth McIntosh, the wife of the commanding officer. Also present at the 'wake' were five former commanding officers of the *Victorious*, Captain J. A. Grindle, Admiral Sir Michael Denny, Captain F. B. K. Stevens, Rear-Admiral Dickenson and Rear-Admiral Janvrin, for whom it must have been a poignant day.

The carrier herself remained in D Lock until Tuesday 5 December, when she was towed back round to Middle Slip Jetty and destoring commenced. During that month, with most of the ship's company still intact, Wren Carmen Brykiert, who was 'Miss Fleet Air Arm' that year, helped Captain McIntosh to stir the Christmas pudding and on Christmas morning a family carol service was held on board. Time, however, was running out for the *Victorious* and in early 1968 large numbers of the ship's company were drafted to HMS *Hermes*, which was due to commission for the fourth time on 17 May 1968, and which lay just a few hundred yards away from the *Victorious* at South Railway Jetty. The ship remained alongside in the dockyard whilst destoring and de-equipping continued, and she was finally decommissioned on Wednesday 13 March 1968. That evening the traditional ceremony of lowering the Colours at sunset was performed for the last time. Officers and ratings who were still on board, together with press represenatives and other guests, gathered near the ensign staff to witness the hauling down of the Colours and to pay a last formal mark of respect to the 'Mighty *Vic*'. Two Royal Marine buglers who had served in the ship's detachment sounded 'Sunset', and the White Ensign and the commissioning pennant were struck for the last time, with the latter being presented to Captain McIntosh, who was to be promoted Rear-Admiral in the following month. The occasion was highly emotional and as the last notes of the bugles died away the silence was shattered by the roar of approaching aircraft, as out of the evening sky from the direction of the Isle of Wight roared three Buccaneers and three Sea Vixens flying over their mother ship in a final salute. They represented the last two strike squadrons which had operated from the ship, 801 and 893, and their arrival was precisely timed for the end of the ceremony.

Soon after this the *Victorious* was towed out into Fareham Creek and she was moored off Elson to await disposal. Apart from shipbreakers the only other commercial interest in the ship came, ironically, from a group of German businessmen who were interested in turning her into a floating exhibition hall. They visited the old carrier twice in the New Year of 1969 and it was reported that they had made an offer for her which was greater than the scrap value. It seems that they were interested in making her seaworthy once again, which would have been a reasonably easy task, and then sending the old carrier on worldwide voyages to exhibit German goods. However, the plan did not come to anything and in July 1969 it was announced that she had been sold to the Scottish Shipbreaking Industries Yard and that she would leave Portsmouth on Friday 11 July 1969. It was at about 4.30pm that day when the deserted aircraft carrier was towed out of harbour and it was a sad sight, with no sailors lining the flight deck, no aircraft on board, and no ensigns flying. With the controversy about the withdrawal of the aircraft carriers still very much alive, the Navy played down the farewells but there was one poignant gesture to mark the aircraft carrier which had served the Navy and the country so well. As the *Victorious* made her way slowly down harbour, with the tugs *Griper*, *Capable*, *Boxer*, *Forceful*, *Samson* and *Dalmation* shepherding her, the last serviceable Swordfish bi-plane flew farewell circuits over her. Piloting the aircraft was Lt-Cdr Trounson RN, from RNAS Yeovilton, while in the observer's seat was Vice-Admiral Sir Richard Janvrin, who was the Flag Officer Naval Air Command. Escorting the Swordfish was a Wessex helicopter and as the 'old lady' reached The Point, holidaymakers and other sightseers crowded Old Portsmouth, the Round Tower and Sally Port to watch the *Victorious* depart on her final voyage. At the Outer Spit Buoy the dockyard tugs handed over to three ocean-going vessels, *Reward*, *Agile* and *Samsonia*, with the towing master, Captain R. F. Dunkley, aboard the latter tug.

The *Victorious* arrived in the Gare Loch five days later, and after another 12 months had passed she had almost ceased to exist. It is ironical that, just over 30 years later, politicians are seriously considering plans for two new aircraft carriers for the Royal Navy. A belated recognition of the power and flexibility of naval air power.

Vice-Admiral Janvrin, in the observer's seat of the last serviceable Swordfish aircraft, gives the final salute as the *Victorious* is towed from Portsmouth Harbour. (*Imperial War Museum*)

Appendix One

Principal Particulars

Length Overall:	1941: 753ft
	1958: 781ft
Beam Overall:	1941: 95ft - 9in
	1958: 103ft - 6in
Standard Displacement:	1941: 23,000 tons
	1958: 30,530 tons
Draught:	1941: 28ft - 6in
	1958: 31ft
Armament:	1941: 16 x 4.5in HA/LA guns in eight twin mountings. Six multiple pom-poms.
	1958: 12 x 3in
Aircraft:	1941: 30 torpedo-bombers; Six fighters.
	1958: 28 - 30 aircraft. Could operate up to 54.
Main Propulsion Machinery:	1941: Three shaft: Three sets Parsons geared turbines. Steam provided by six Admiralty three-drum boilers in three boiler rooms.
	1958: Three shaft: Three sets Parsons geared turbines. Steam provided by six Foster-Wheeler boilers in three boiler rooms.
SHP:	111,000 SHP
Speed:	30 knots
Complement:	1941: 1,286
	1958: 2,400
Flight Deck:	1941: Axial flight deck
	1958: $8^{3}/_{4}°$, fully-angled flight deck
Deck Recognition Letter:	1958: V
Pennant Number:	R 38

Appendix Two

Commanding Officers HMS *Victorious*

	Date Appointed:
Captain H. C. Bovell CBE RN	15 October 1940
Captain L. D. Mackintosh DSO DSC RN	23 November 1942
Commander R. C. V. Ross DSO RN	1 November 1943
Captain M. M. Denny CB CBE RN	8 December 1943
Captain J. C. Annesley DSO RN	15 August 1945
Captain E. B. K. Stevens DSO DSC RN	30 September 1947
Captain N. V. Dickinson DSO DSC RN	25 May 1948
Captain J. A. Grindle CBE RN	16 July 1949
Captain C. P. Coke DSO RN	16 December 1957
Captain H. R. B. Janvrin DSC RN	11 August 1959
Captain J. M. D. Gray OBE RN	21 December 1960
Captain P. M. Compston RN	19 November 1962
Captain D. L. Davenport OBE RN	6 October 1964
Captain I. L. McIntosh DSO DSC MBE	12 September 1966

Appendix Three

Former Ships

The First *Victorious*
A third-rate (74 guns), she was launched on 27 April 1785 at Blackwall on the River Thames at a time when Britain was at peace, having established peace terms with the newly independent United States. However, between July and September 1795, when Britain declared war on the Dutch, she was present at the capture of the Dutch outpost at the Cape of Good Hope. This was one of the most important ports on the route to India, which was being developed following the loss of the American colonies. On 9 September 1796 she was in action against a French squadron in the Strait of Malacca, but seven years later she was sold and demolished at Lisbon.

The Second *Victorious*
Built on the Beaulieu River in the New Forest, the next *Victorious* was another third-rate (74 guns) of 1,746 tons, and she was launched at Bucklers Hard in October 1808, when the country was locked into the Napoleonic Wars. Ten months after her launch, under the command of Captain G. E. Hammond RN, she took part in the Walcheron Expedition where British troops were landed on the islands of Walcheron and South Beveland on the River Schelde to prepare for the capture of Antwerp. The intention was to hold Walcheron and to keep it open for British trade, but in the marshy conditions which prevailed the troops were decimated by fever and in December that year Walcheron was evacuated. Far from opening a second front, it had been a costly mistake. Three years later, on 22 February 1812, under the command of Captain J. Talbot RN, the *Victorious*, together with the brig *Weazel*, was at Venice under orders to destroy a new French warship, *Rivoli*, which was fitting out at the port. In the event the British warships sank the French warship *Mercure*, damaged the *Jena*, and after an action lasting five hours forced the severely damaged *Rivoli* to surrender, after which she was added to the Royal Navy's fleet. The second *Victorious* was demolished in 1861.

The Third *Victorious*
A Majestic-class, pre-Dreadnought battleship with a displacement tonnage of 14,820, and fitted with steam reciprocating engines. She was armed with 16 guns, was launched at Chatham and completed at a cost of £885,212. She commissioned under the command of Captain E. S. Poe RN on 8 June 1897 for service with the Mediterranean Fleet. In February 1898 she was ordered to join the China Station, and on 14 February she ran aground near Port Said. After being lightened she was refloated and went on to make her southbound transit of the Suez Canal. Following her service east of Suez she was recommissioned on 16 May 1900 at Malta, and in February 1904 she was the flagship of Rear-Admiral The Hon H. Lambton, Flag Officer, Second in Command, Channel Fleet. She remained in the Channel Fleet until January 1907, when she became part of the Nore Division, Home Fleet based at Chatham. In March 1909 she was refitted to burn oil fuel and she had new fire control and radio systems added, and in the following month she was attached to the Third Division of the Home Fleet, but remained at her Chatham base. In June 1910, in thick fog, she was involved in a collision with her sister ship, HMS *Majestic*, as a result of which she sustained damage to her sternwalk and her starboard engine room.

From January 1911 to May 1912 she formed part of the Devonport Division, before becoming part of the 3rd Fleet. In December 1913 she underwent a short refit in Devonport Dockyard and she was earmarked to join the 9th Battle Squadron in August 1914, but on the outbreak of war, being hopelessly outdated, she was sent to the River Humber to act as the guardship. In February 1915 she was sent to Elswick on Tyne to be disarmed, and her 12-inch turrets were fitted in the new monitors, *Prince Rupert* and *General Wolfe*. In March 1916 she was converted to a dockyard repair ship and she was based at Scapa Flow where she flew the flag of the Vice-Admiral Orkney & Shetlands. In 1920 she was renamed *Indus II* and in December 1922 she was sold for demolition at Dover.

The Fifth *Victorious*
The fifth, and current, Royal Navy warship to bear the name *Victorious* is the Vanguard-class nuclear-powered ballistic missile submarine which was first commissioned in 1995. She displaces some 15,000 tons when dived, and she has an overall length of 150 metres, with a breadth of 13 metres. Her steam turbines give her an underwater speed in excess of 25 knots and she is armed with 16, Trident 2 missiles, although she also retains the capability of firing conventional torpedoes. The UK's strategic nuclear deterrent is primarily vested in the Vanguard-class submarines, and in order to allow them to remain at sea for long periods each vessel has two crews, port and starboard, of 135 officers and ratings.

Appendix Four

HMS *VICTORIOUS*

A Guide to Civilian Passengers

Section I.
The ship's company of HMS VICTORIOUS will do their best to make your passage to the United Kingdom an enjoyable one. At the same time you must remember you are not in a Liner, but are privileged to be a passenger in one of His Majesty's Ships. Life on board must continue to be governed by service rules and customs and we ask you, therefore, to comply as far as possible with our Service Routine.

You will find it difficult during the first few days to find your way around; do not hesitate to ask any one of us should you get lost, we will be glad to help.

Here, for your information, are some Rules and Service Customs with which you should conform.
(1) Stand up and make way for the Captain and any Officer when they are going 'Rounds' officially. The 'Rounds' take place every evening at 20.45 and at 11.00 on Saturdays.
(2) Help keep the ship clean and tidy. Place cigarette ends, matches, etc, into receptacles provided. Never throw things out of scuttles or over the ship's side. Take care not to damage the paintwork.
(3) Do not get in the way of or distract men who are on duty. You are allowed nearly everywhere in the ship except for men's living spaces, but work must not be interrupted.
(4) Comply with the ship's regulations for smoking. You must not smoke during working hours except in your own sitting rooms and quarters.

Passenger Enquiry Office.
There is an Enquiry Office in the Island Superstructure which is open from 09.00 to 12.00 and from 13.15 to 16.00 (ie 1.15 to 4pm), and every effort will be made to help you in your problems.

Welfare.
An officer is carried in the ship who is a trained Officer in Welfare Activities. He can be of considerable assistance in matters concerning the relationship of wives and the Services. The Officer, Sub Lieutenant Whitehead RNVR, may be contacted through the Enquiry Office. Mrs Fordham, the Lady Almoner, is also available to assist.

Church Services.
Church Services are held each Sunday and on weekdays. The times of service and the places in which they are held are published in Commander's Orders and on the Church Notice Board.

Chapel.
There is a Chapel on board which is open for prayer at all times. The ship's Chaplain is The Rev H. D. Halsey RNVR, who has a cabin next to No 84, and will be glad to give any help that he is able.

Shop.
There is a small shop in the Wardroom Lounge at which it is possible to buy sweets, cigarettes, etc. Ice cream and soft drinks may be bought at the Canteen on the Flight Deck.

Library.
The women passengers' library, Lower Library is situated in the Upper level adjacent to the chapel and the Torpedo Body Room. Times of opening and rules to be observed are promulagated on a separate notice.

Cinema.
Shows are held daily, times of which will be promulgated.

Hairdressing.
   A small Hairdressing Salon has been made in the after end of the lounge adjacent to 'U' Dormitory (Upper Gallery Deck). It will be staffed by volunteers from among the passengers. Times of opening, scale of charges etc, will be promulgated separately.

Bathing and Lavatory Facilities.
   In all warships, these facilities are usually grouped together in spaces not needed for armament, engines, etc, and it must be admitted that they are not always in the most convenient places. We have increased these facilities as much as possible; showers and lavatories are to be found as follows:-
   Passengers in Cabin Flats and Dormitories Q, R, S and T (Port Hangar Deck).
   Lavatories are right aft on the Upper Deck.
   Bathroom on the Upper Deck forward of the Wardroom Flat.

Hangar Dormitories G-P inclusive.
Shower Room: At the fore end of the Hangar, starboard side.
   Lavatories: Leave the Hangar by the centre door, starboard side, turn either right or left along the passage and go up two ladders. There you will find two groups of lavatories clearly marked. There are also 'Elsan' closets outboard of each Dormitory.

Recreation Space Deck Dormitories U, V and W.
   Shower rooms and lavatories are situated in the Rec Space.

Upper Gallery Deck Dormitories X, Y, and Z.
   Shower rooms between X and Y Dormitories. Lavatories aft of Z Dormitory.

Laundry.
   Limited facilities only exist for laundry onboard. Laundry must be carried out by each individual concerned. The laundry is to be found right forward, port side of the Hangar, and clothes may be hung on the lines provided in the forward lift well, or in 'C' Hangar during bad weather.
   Ironing boards and electric irons are provided in each Dormitory.

Handicrafts.
   Materials for leathercraft and felt toy making can be obtained from Mr Greenwood, the Senior Master, in the School Room which is situated in the Island Superstructure one deck up from the Flight Deck.

Money and Valuables.
   Safe deposit is available on board in which to place money, jewellery, etc. Passengers wishing to deposit valuables should provide themselves with strong envelopes in which to place the articles and a receipt will be given.
   Passengers are reminded that Travellers' Cheques cannot be cashed on board.

Baggage.
   Due to the limited stowage space in Dormitories, it is possible to allow only one suitcase and one drawer per person.
   The remainder of your baggage will be placed in the fore end of 'C' Hangar.
   The baggage will be stowed alphabetically and will be available to you every forenoon between the hours of 09.30-11.30.
   All queries referring to baggage will be dealt with by the Baggage Officer, Lieutenant Griffin RNR.

Women's Service Officers.
   Eight women have volunteered and are on board to advise and help you and to assist in the organisation of the journey. Each of these Women's Service Officers has a group of dormitories under her special care and will always be available to help you.
   The Duty Women's Service Officer will sleep in U dormitory (Rec Space) and will go Rounds during the night.

Section II

THE FOLLOWING REGULATIONS HOLD GOOD ABOARD ALL HM SHIPS

(1) Every person aboard one of HM Ships is subject to the Naval Discipline Act.
(2) It is contrary to the King's Regulations and Admiralty Instructions to bring on board, have in one's possession, or consume any Spirits, Wine or Beer, except for the Tot of Rum issued to those entitled.
(3) Gambling in any form is forbidden.
(4) No unauthorised person is permitted to interfere with Ship's Fittings, Armament, or Boats etc.
(5) In the event of 'Fire' it is the duty of the person detecting it, firstly to attempt to put it out, and secondly to inform the Officer of the Watch and Damage Control Headquarters. Details of the firefighting organisation are published separately.

Section III

INSTRUCTIONS FOR WOMEN PASSENGERS.

(1) In the event of Fire or any Emergency, assistance may be obtained by informing the Duty Women's Service Officer, the Baggage Sentry, or by pressing the Fire and Alarm bells situated in all dormitories and Hangar Access lobbies.

(2) The following are out of bounds to Women:-
   (a) Ratings living spaces and messes.*
   (b) Officers' cabins and messes.*
   (c) Below the level of the Hangar Deck, except in the after part of the ship where they are allowed on the Upper Deck as far forward as the forward end of the Main Baggage Store (Bulkhead 88) and down to the Women's Cabin Flats.
   (d) Above the level of the Education Centre in the Island, ie, one deck up above the Flight Deck.
   (e) Gun positions and Galleries and inside boats.
   (f) Women's Service Officers may go into Officers' Messes if invited by a ship's officer.

NOTE*:- Except for organised entertainments, when permission is first to be obtained from the Commander.

(3) The Ship's Company are not allowed in the Women's Quarters and living spaces except as required for duty. Permission may be obtained from the Commander in cases of organised entertainment.

Suggestions and Complaints:-
   Any suggestions for improving the comfort and welfare of Passengers will always be welcomed. Suggestions should be handed in on a chit to the Passengers' Enquiry Office, which is situated in the Island.
   Massed complaints are not allowed. If there is a general complaint about some matter, the Duty Women's Service Officer should be informed, and she will bring the matter to the notice of one of the following Officers:-
   (a) Lt Cdr Pellow RN.
   (b) Lt Cdr Pollock RN.
If the matter is urgent, one of these Officers may be approached direct.
(5) Be careful never to waste water. This is to be stringently rationed at sea and if you have more than your share someone will have to go without.
(6) Never hang anything out of a scuttle or port hole or anywhere where it can be seen from outboard.
(7) Women may use the Flight Deck in addition to their own quarters for smoking and recreation, provided they do not interfere with work in progress and do not smoke during working hours except in authorised places. They are never to smoke near Aircraft at any time.

(8) Leave:-
It must be remembered that if you were in a troopship you would get no leave in any port. There will probably be a chance to land at Colombo and, perhaps, at Port Said (if the international situation allows), but leave can not be looked upon as certain. Any leave given will be broadcast and published on all Notice Boards. Persons failing to return to the ship by the stated time are liable to be left behind.

GENERAL NOTES

(1) DINING ARRANGEMENTS.

All Brides will take their meals in the Dining Hall in B Hangar. There will be two sittings, and meals will be served at the following times:-

```
BREAKFAST.      1st  ..........07.30 - 08.00
                2nd  ..........08.30 - 09.00

DINNER.         1st  ..........11.30 - 12.00
                2nd  ..........12.30 - 13.00

AFTERNOON TEA.       ..........15.15 - 15.45

SUPPER          1st  ..........18.30 - 18.45
                2nd  ..........19.30 - 19.45
```

For the first week, Brides accommodated in Q, R and Hangar Dormitories will take their meals first sitting, the remainder of the Brides second sitting; thereafter the sittings being changed round weekly, commencing with breakfast on Mondays.

(2) SLEEPING BERTHS
Sleeping accommodation consists as follows:-

Unit	Type	Letter	No of Berths	
(1)	Single and Double Cabins	Nil	58 berths	
(2)	Officers' Dormitories	Nil	38 berths	
(3)	Camp Beds in Cabins	Nil	26 berths	
(4)	Dormitory Accommodation in Hangar	I to P	8 dormitories	36 berths
(5)	Enclosed Dormitories in Port Hangar Deck	Q to R	2 dormitories	16 berths each
		S & T	2 dormitories	12 berths each
(6)	Recreation Space Dormitory	U	126 berths	
(7)	Enclosed dormitories stbd side Recreation Space	V	20 berths	
(8)	Enclosed dormitory port side Recreation Space	W	16 berths	
(9)	Enclosed dormitory 79 Mess	X	12 berths	
(10)	Enclosed dormitory 80 Mess	Y	12 berths	
(11)	Enclosed dormitory 81 Mess	Z	16 berths	

```
        Total in Unit (4) .............320 berths
        Total in Unit (5) .............56  berths
        Total in Units(6) to (11) ........202 berths
        Grand Total  ................700 berths
```

It is appreciated that the berths vary in comfort and location, but every effort has been made to allocate them as fairly as possible.

Appendix Five

HMS *Victorious*

Battle Honours:

Rivoli 1812
Bismarck 1941
Norway 1941-44
Arctic 1941-42
Malta Convoys 1942
North Africa 1942
Biscay 1942
Sabang 1944
Palembang 1945
Okinawa 1945
Japan 1945

Appendix Six

Aircraft Types - HMS *Victorious*

Fairey Swordfish: Three-seat torpedo-bomber reconnaissance biplane. First flown April 1933, entered service FAA July 1936. Powered by 750hp Bristol Pegasus radial piston engine, 86knots, 322 nautical miles. One forward-firing synchronised .303 machine-gun in forward fuselage & one .303 Lewis gun in rear cockpit. One 610lb torpedo or depth charges or bombs up to 1,000lb.

Fairey Fulmar: Two-seat carrier-borne fighter. First flown January 1940, entered service FAA July 1940. Powered by 1,080hp Rolls Royce Merlin VIII piston engine, 230 knots, 809 nautical miles. Armament eight .303 machine-guns in wings.

Hawker Sea Hurricane: Single-seat carrier-borne fighter. Entered service FAA 1941. Powered by 1,280hp Rolls Royce Merlin XX piston engine, 342mph, 460 miles. Four 20mm forward-firing cannon.

Fairey Albacore: Three-seat torpedo-bomber biplane. First flown December 1939, entered service FAA March 1940. One 1,130hp Bristol Taurus XII radial piston engine, 101 knots, 809 nautical miles. One forward-firing .303 machine-gun in stbd wing, twin .303 machine-guns in rear cockpit. One 1,610lb torpedo, six 250lb or four 500lb bombs beneath wings.

Fairey Barracuda: Three-seat torpedo- and dive-bomber. First flown December 1940, entered service FAA January 1943. One 1,640hp Rolls Royce Merlin 32 piston engine, 149 knots, 417 nautical miles. Two .303 machine-guns, plus one 1,620lb torpedo or up to 1,600lb of bombs, or six 250lb depth charges, or 1,640lb of mines.

Grumman Martlet: Single-seat carrier-borne fighter. Entered service FAA Sept 1941. Twin Wasp SC34-G piston engine, 280 knots. Four .50mm Colt machine-guns.

Chance-Vought Corsair: Single-seat fighter-bomber. Entered service FAA 1943. One R2800-8 piston engine, 361 knots, 726 nautical miles. Six .50mm Colt machine-guns, one 1,000lb bomb.

Grumman Avenger: (Also called Tarpon in UK) Three-seat torpedo-bomber. Entered service FAA 1944. One Wright Cyclone R-2600-20 piston engine, 131 knots, 982 nautical miles. 2,000lb of bombs, one 1,620lb torpedo, depth charges.

Supermarine Walrus: Four-seat spotter-reconnaissance or air-sea rescue seaplane. First flown June 1933. One 775hp Bristol Pegasus VI radial piston engine, 235mph, 600 miles. One Lewis gun in bow position and two amidships, plus 760 lb of bombs/depth charges on underwing racks.

de Havilland Sea Venom: Two-seat carrier-borne all-weather strike aircraft. First flown 1954, entered service FAA May 1955. One de Havilland Ghost jet engine, 547 knots, two hours' endurance. Four 20mm Aden cannon, provision for bombs and rocket projectiles.

Supermarine Scimitar FI: Single-seat carrier-borne fighter-bomber. First flown April 1954, entered service FAA August 1957. Two Rolls Royce Avon 202 turbojets, 617 knots, 1,500 miles. Four 30mm cannon or four AIM-9B Sidewinder, or 12, 2-inch rockets. Four 500lb or two 1,000lb bombs. Four Bullpup missiles. Nuclear capability.

de Havilland Sea Vixen FAWI: Two-seat all-weather interception and strike aircraft. First flown Sept 1951, entered service FAA July 1959. Two Rolls Royce Avon 208 turbojets, 560 knots, 1,200 miles. Four Firestreak air-to-air missiles, or four rocket packs, plus 28-51mm rockets. Two 1,000lb iron bombs.

Blackburn Buccaneer SI: Two-seat shipborne low-level strike aircraft. First flown April 1958, entered service FAA March 1961. Two de Havilland Gyron Junior 101 turbojets, 626 knots, 1,200 miles. One 4,000lb internally carried bomb, four 1,000lb external underwing bombs. Nuclear capability.

Douglas Skyraider: Three-seat airborne early-warning aircraft. Entered service FAA October 1951. One Wright Double Cyclone R-3350-26WA piston engine, 305 knots, endurance 7 hours. Equipped with AN/APS 20A pulse search radar.

Fairey Gannet AEW 3: Three-seat, airborne early-warning aircraft. Entered service FAA August 1958. One Double Mamba 102 piston engine, 217 knots, endurance six hours. Equipped with AN/APS 20F radar.

Westland Whirlwind HAS1: Shipborne ASW helicopter. Three crew (pilot, observer, sonar operator or SAR swimmer when in planeguard role). One Alvis Leonides Major 775 piston engine, 95 knots. One 53cm torpedo or depth charge. Equipped with AN/AQS-4 dipping sonar.

Westland Wessex HAS 7: Shipborne ASW helicopter. Three crew (pilot, observer, sonar operator or SAR swimmer when in planeguard role). One Napier Gazelle 160 turboshaft engine, 120 knots. Two Mk 44 torpedoes, two Mk II depth charges. Equipped with Type 195 dipping sonar.

Acknowledgements:

My thanks to: Vice-Admiral Sir Ian McIntosh KBE DSO DSC, the *Victorious'* last commanding officer, for kindly writing the foreword to this book: Mr Brian Conroy for his magnificent watercolour painting on the front cover of the book: Mr John S. Morris for his pen and ink sketches used on the endpapers.

I must also thank the following for their help and, in many cases, for the loan of very valuable photographs:-

Jim Allaway, Editor, *Navy News:* Raymond Barker, Hereford, for permission to quote from his book '*Victorious* The World Over': Roger Beacham and the staff of Cheltenham Reference Library: James Bolton, Norwich: Ian Carter, Dept of Photographs, Imperial War Museum, London: Michael Cassar, Valletta, Malta: Lieutenant Norman Curnow RN (Retd), Helston, Cornwall: Barry Elliott, Chandlers Ford, Hampshire: Bernard Elliott, Blyth, Northumberland: Carol Farr, *The News*, Portsmouth: Stanley Filmer, Walton-on-the-Naze, Essex: John Harper, Coventry: Charles Heath, Waterlooville, Hampshire: R. Hobbs, Shanklin, Isle of Wight: David Hooper, Rochester: Thomas Hopwood, Port Glasgow: Michael Lennon, Waterlooville, Hampshire: Bernard Long, Hong Kong Government Information Service: Charles Minett, Cheltenham, Gloucestershire: Dr Mark Nicholls, Cambridge University Library: Captain Hugh Owen RN (Retd), Chichester: Anthony J. Perrett, Gosport, Hampshire: J. P. Phillimore, Poole, Dorset: David Richardson and Jerry Shore, Research Officers, Fleet Air Arm Museum, Yeovil, Somerset: Ian Spashett, FotoFlite, Kent, for permission to use photographs from his company's collection: R. J. Stapleton, Yeovil, Somerset: Alexander Thomas, Motherwell: Leslie Vancura, Killay, Swansea: Special thanks to my wife Freda and my two daughters Caroline and Louise for their invaluable help.

Also In The Series

HMS *Eagle* 1942-1978 £18.95
(Plus £2.00 p&p in UK/EU or £4.00 for airmail to all other countries)

HMS *Centaur* 1943-1972 £16.95
(Plus £2.00 p&p in UK/EU or £4.00 for airmail to all other countries)

Other Titles From FAN PUBLICATIONS
SS *Canberra* 1957-1997 £21.00
(Plus £2.00 p&p in UK/EU or £4.00 for airmail to all other countries)

Famous British Liners Series.
For current list write to:

FAN PUBLICATIONS
17 Wymans Lane
Cheltenham
Glos GL51 9QA
England
Tel/Fax: 01242 580290